U0944297

顶级设计空间
TOP DESIGN SPACE
《顶级设计空间》编委会 编

创意办公

CREATIVE OFFICE

大中型办公 & 工作室

LARGE OFFICE & STUDIO

（第二版）

中国林业出版社

创意办公：汉英对照 / 《顶级设计空间》编委会编
. -- 2版. -- 北京：中国林业出版社, 2013.4
（顶级设计空间）

ISBN 978-7-5038-7274-7

Ⅰ. ①创… Ⅱ. ①顶… Ⅲ. ①办公建筑－室内装饰设计－图集 Ⅳ. ①TU243-64

中国版本图书馆CIP数据核字(2013)第274862号

《顶 级 设 计 空 间》 编 委 会 编

主编：张青萍
编委：孔新民、贾陈陈、许科、李钢、吴韵、竺智、曾丽娴

责任编辑：纪亮
英文翻译：董君、梅建平、牛晓霆、万毅、赵强

出版：中国林业出版社
（100009 北京西城区德内大街刘海胡同 7 号）
网址：http://lycb.forestry.gov.cn/
E-mail：cfphz@public.bta.net.cn
电话：（010）83143581
发行：新华书店
印刷：北京卡乐富印刷有限公司
版次：2016年4月第2版
印次：2016年4月第1次
开本：230mm × 300mm
印张：32
字数：360千字
定价：199.00元

设计进行时

中国改革开放30年，室内设计行业行进到今天，也已经有了无计其数的变化、发展和积累，30年的思考、30年的实践、30年的进步，也造就了这30年的成绩。

我们好似在进行着一场接力赛，祖先把中华民族灿烂的文化一代代地传承到21世纪，我们有责任将这份优秀的遗产倍加珍藏以传给后代。我们所面临的挑战是拿什么当代的室内文化馈赠后人？但其实这30年中由于电子技术的普及和信息的迅速传播和交换，设计已出现国际化、同一化的倾向，与此同时引起了传统性、地域性和个性差异的不断丧失，又有由于社会追求物质与功能价值的同时造成对精神和文化价值的忽视，我们已找不到回头的路。但不管历史结论会如何，我们这代人是努力的、勤奋的，是不断地用自己的智慧为中国室内设计行业进步奉献着的。

总的看来，21世纪的室内设计发展有以下倾向和趋势：

倡导绿色设计

人类起源于自然，其间虽曾摆脱过自然，但最终还是要以全新的面目去回归于自然。如此轮回恰恰历史地、辩证地道出了人与自然关系的变化。如今的人们，特别是生活在大城市里的人离大自然已越来越远了，于是人们特别希望在室内再现一些大自然的情景，以求得哪怕是暂时地、局部地享受。作为设计师一方面尽可能地创造出生态环境，让人们最大限度地接近自然，另一方面须有环保意识，努力去提高设计中的健康因素，以满足人们在生理和心理上的需要。

室内设计中的健康设计充分利用自然或仿自然的因素，为人们提供生活舒适的空间。室内的色彩、照明及功能空间的弹性分割，都应该在满足其基本功能的基础上，尽可能充分利用自然能源。尤其是提倡对绿色装饰材料和绿色照明材料的运用，同时注重社会心理学的研究。绿色设计本着以下几个原则：

①设计上使用最少的材料，少浪费应节约的资源、能源，力戒奢靡；②尽量多采用污染少，环保性能强，安全可靠的材料；③符合人体工程学的要求，讲求空间上比例与尺度，避免使人感到压抑或繁琐；④设计应以人为本，满足人的本质需要。

运用高科技手段

科技进步影响着人类生活的方方面面，现代科技的发展为人类的衣食住行提供了很多方便，它可以使居住环境更加符合人们的意愿。因此，在设计中提高科技含量，创造高效率、高功能、高质量室内生活环境的要求已愈加鲜明。在本世纪，楼宇的智能化将逐步实现。建筑智能化就是将结构、设备、服务运营及相互关系进行全面综合配置，从而达到最佳的组合程度，使建筑具有高效率、高功能和舒适性。单从自动化来看，就是实现建筑设备自动化、办公自动化和通讯自动化。建筑智能化并非仅仅针对大楼宇，还会迅速走向寻常百姓家。住宅中自动防盗报警，自动调温、调湿、自动除尘、调节灯光亮度、自动控制炊事用具等，如今也已成为事实。

注重设计的系统性

在这个日新月异、急剧变化的时代，系统地看待问题和解决问题是当代人的特质。站在现代与历史之间的人们既希望从传统中找回精神的家园，以弥补快速发展带来的心理失落与不安；同时又满怀着激情和野心试图运用当代技术和审美重新诠释历史，使之适应现代生活。系统的眼光可使我们将面临的室内设计浅层次问题渗透到更深的层次方面加以科学、综合地解决。这样的设计系统性包括环境学、生态学、经济学、系统论、方法论、控制论、统筹学、管理学及有关室内设计方面的政策法规、标准规范等方面的内容。

室内设计系统是指应用系统的观点和方法，将室内设计的内容、要素，相关的领域和环节，以及室内设计的程序予以统筹而形成的一个框架体系。从与其相关部分的关系和进行的程序来分析，可理解为有横向设计系统和纵向设计系统两个方面。横向系统设计表现为在设计过程中所涉及到的如生理学、心理学、行为科学、人体工程学、材料学、声学、光学、经济学等诸多因素；纵向系统设计表现为对设计实现过程中所有历程的考虑。概括而言，横向系统设计强调相关与联系；纵向系统设计强调过程与变化。

无论将来的室内设计如何发展，它都必然在一个更光泛、更全面的系统里科学地伸展，设计师也必将持有这种科学的态度和掌握这类理性的方法而从事设计。

本系列图书刊登的是近年来一些顶级的设计作品，它们或多或少的反映的是当下室内设计师的思考和当前技术背景下的实践。30年告了一个段落，下一个30年又将开始，我们已走上新的征程，设计永远是进行时。愿这本套书的出版能得到业界的认可和赞扬！

南京林业大学 风景园林学院副院长、教授

id+c《室内设计与装修》杂志主编 张青萍

2010.3.1草于南京

大中型办公

工作室

Large office

Studio

UNIVERSAL EXPRESS

宇宙运通国际纺织公司

01

【坐落地点】上海静安区延平路98号；【面积】3F面积 950 m²，4F面积 950 m²
【设计】赖建安；【设计公司】十方圆设计工程公司
【主要建材】浇注清水混凝土、条纹玻璃、流沙墙、锈铁板、水曲柳染色、柚木实木
【摄影】深蓝

设计师采用一个具东方LOFT的SOHO空间：以东方玄学太极动点的概念融合建筑尺度，强调回归自然本质、创造空间比例，颠覆旧有审美角度，升华出禅味的哲学。

设计团队把动线归整到最少，来贯穿全区，并且推敲如何创造舒适的空间比例。在洽谈室的塑造手法上用了箱体空间，设计师把这个箱体设计确定在动线节点上，使创造出来的空间有内外之分，以期待在单调无趣的办公空间中增添点乐趣。设计师及其团队希望设计元素在视觉上、动点处理上、触觉上，以及诸多设计元素造成的相互影响与融合上，使得原本传统而呆板的办公室环境产生相当浓厚的艺术氛围，营造出更自然舒适的空间，让客户与员工在这空间中能感觉放松。

Designers use a LOFT in SOHO space with the East: The Oriental concept of Taiji integration of architecture, emphasizing the nature of return to nature, and create spatial scale, to subvert the old aesthetic point of view, sublimation of a Buddhist Feeling philosophy.

Design team to move the entire line go to a minimum, to run through the region and to refine how to create a comfortable space for the ratio. In negotiating room used to shape the way the cabinet space, designers designed this box to identify moving line of nodes, so that the space created inside and outside of the points, in anticipation of office space in the monotonous and uninteresting to add a little fun . Designer and his team hope that in the visual design elements, the fixed-point handling, tactile, as well as a number of design elements created by the mutual influence and integration, making the original traditional and rigid office environment quite a strong arts culture, and create a more natural and comfortable space for customers and employees in this space can feel relaxed.

↑会议室的布置 / The conference room layout.

↑入口接待处，接待台的设计很有特点 / Entrance reception area, reception stations designed with unique features.

→传统的木门作立面的装饰 / To make the traditional wooden facade decoration.

↓进入公司，全玻璃的独立空间增加了空间的变化 / Into the company, all-glass room to increase the independence of the space changes.

↑二楼的镂空楼梯，流沙般质感的墙 / On the second floor of the hollow stairs, sand-like texture of the wall.
↓主管办公室手稿 / Director of Office of the hand drawing.

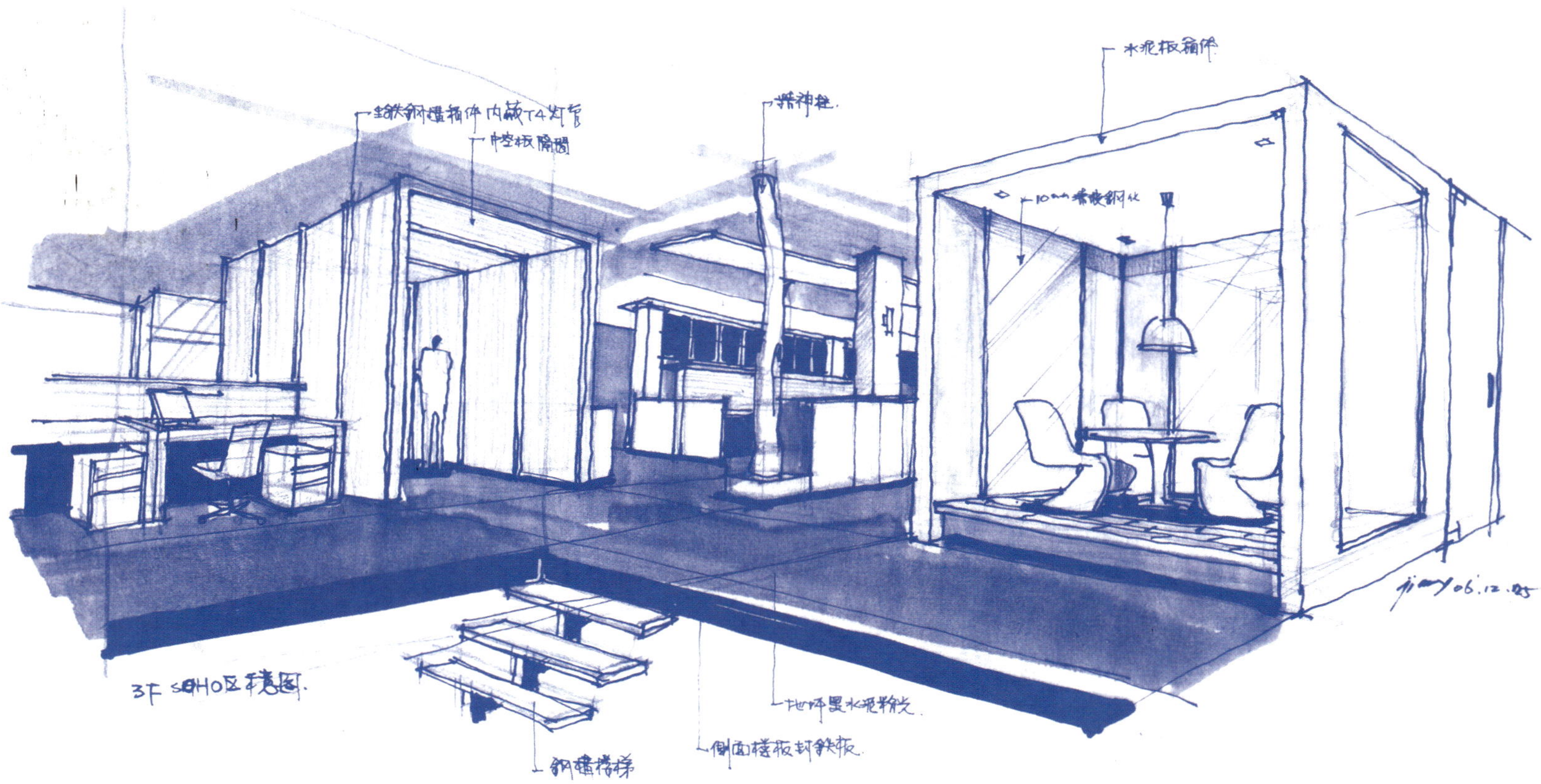

↑ 主管办公室实景，选用了厚重的传统家具 / Virtual Office of the Head, selected heavy traditional furniture.
↓典型的“混搭”效果 / A typical "mashup" effect.

↑当水晶灯“遭遇”官帽椅 / When the Crystal Light "encounter" Official Hat Armchair.
← 主管办公室的书柜 / In charge of the office of the bookcase.
↓办公区域走廊 / Regional office corridor.

LG FASHION STUDY CENTER

LG上海办公室

02

【坐落地点】上海浦东张江花园商务办公园区；【面积】1700 m²
【设计】VEP DESIGN
【设计团队】蒋琼耳、吕永中、刘悦、殷之奇
【摄影】吴永长

整个建筑共分三层，包括一个屋顶花园，总面积约为1700 m²。设计师先大胆地将原建筑内部所有的新增隔断拆除，让我们可以清晰地看到原始的空间布局：建筑结构上采用钢筋混凝土框架体系，通过6m × 6m 的柱网模数在平面呈标准的网格状排列，垂直方向上下共有三格。结合现有的空间结构，设计师又独具匠心地在建筑的内部打开了中心区域二、三两层的楼板，开辟出一个垂直的共享空间。

设计之后的室内布局采用了比较规整的中轴对称方式，以镶嵌有楼梯的中庭作为建筑内部的主轴线，让空间易于统一的同时却又不失变化。建筑的一楼是整个研发中心的主要入口区，其左侧是部分办公区域与辅助区域，右侧是接待等候区和综合展示区。沿着正对主入口的楼梯上到二层，其左侧是综合办公区域，右手边是一个"悬浮"的会议室。三楼是分为左侧的屋顶花园与右侧的高层办公区域两部分。

室内以规整有序的建筑结构为三维框架，在立体的框架中设置不同的功能单元（会议、办公、接待等），通过中庭楼梯等多种形式的交通空间将其串联起来。置身在办公空间内部的人们，在感受令人愉悦的内部外部景观的同时，也能感受到室内外花园、屋顶花园与建筑、室内生动的空间关系。

Whole building is divided into three levels, including a roof garden, with a total area of approximately 1700 m². First, the original designer boldly that all new construction within the partition removed, so that we can clearly see the layout of the original space: building the structure using reinforced concrete frame system, through the 6 m × 6 m of the column net modulus in the plane a standard grid was arranged in the vertical direction up and down a total of three cells. Integration of existing spatial structure, designers also originality in the building within the open central area two or three two-story floor, opened up a vertical shared space.

After designing the interior layout of a relatively structured way of axial symmetry, to inlay a staircase in the atrium inside the building as the main axis, so that space for easy-to-uniform changes in the same time they lose. Building on the first floor is the R & D Center's main entrance area, the left is part of the region and supporting the regional office, on the right is the reception waiting area and a comprehensive exhibition area. Being the main entrance along the stairs to the second floor, the left is a comprehensive office area, right-hand side is a "floating" of the conference room. On the third floor is the roof garden is divided into left and right of the high-rise office area in two parts.

Neat and orderly room with the architecture of three-dimensional framework, in the framework of three-dimensional set of different functional units, through the atrium stairs and other forms of transport space will be the series together. Sitting in the office space within the people feel pleasant in the internal and external landscape, while also felt the indoor and outdoor gardens, roof garden and architectural, interior and vivid spatial relations.

↑空中会议室，像一只绿色的半透明盒子 / Air meeting room, like a translucent green box.

↑↑中庭楼梯通透自然，充满时尚感 / Atrium stairs transparent nature, full of fashion sense.
←一层等候区 / Floor waiting area.
↓一层平面图 / Floor plan.

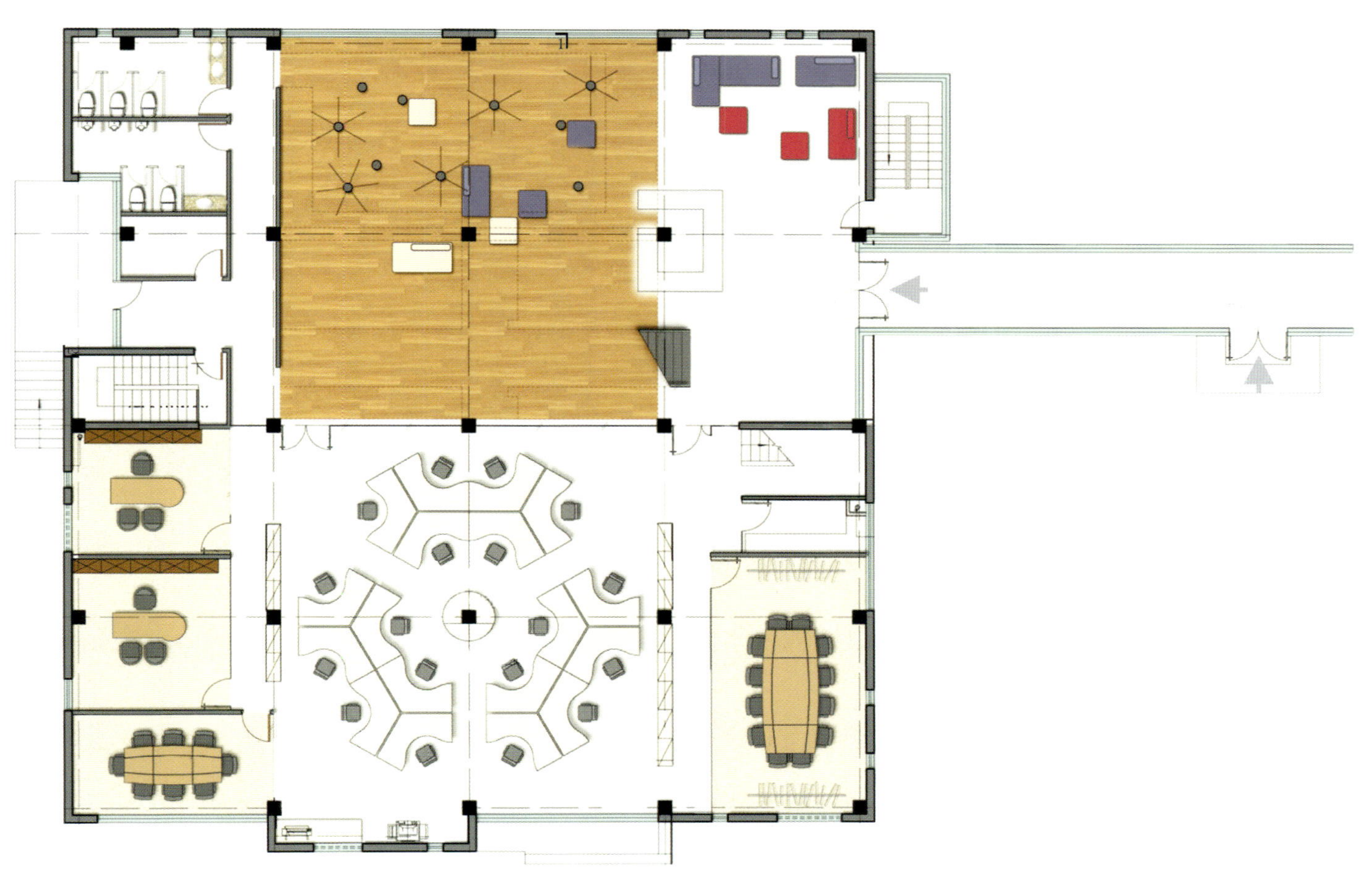

↑楼梯平台 / Staircase platform.
← 服装裁剪区的绿色隔断富有创造性 / Tailoring creative cut off the green zone.
↓二层平面图 / 2nd floor plan.

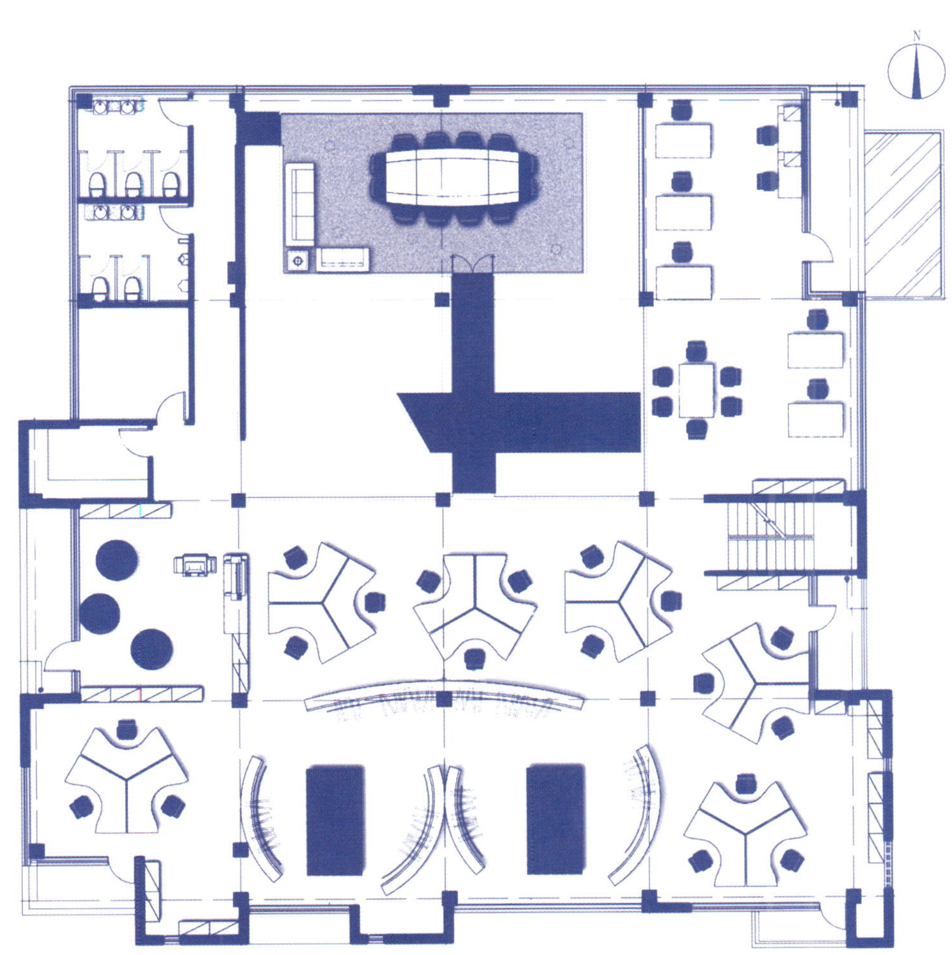

BERGAMO INDUSTRIALISTS UNION

贝加莫工业联盟总部

03

【坐落地点】意大利贝加莫
【设计】Buratti + Battiston Architects
【参与设计】Gabriele Buratti, Ivano Battiston, Oscar Buratti,Massimiliano Gini, Elena Crespi, Roberta Numi
【摄影】Andrea Martiradonna

这是一个改造设计,设计全面地考虑了三个相关联空间序列的过渡与融合，使整体室内空间凝练紧凑，又不乏设计的活泼与灵动。大厅面向外部的公共空间，其入口的标志醒目明显，常用的透明展示橱窗被大量黑色镜面玻璃所代替，反射出城市喧嚣的街景。从门厅进入室内空间，大红色镜面玻璃的热情被轻质木材温暖包裹，各种有趣的插曲在此上演，各种极具表现力的价值观念在此融合。天窗的设计也极具特色，充分利用自然采光，采用四个真正的自然光源，配合四个人造光源，天花板内隐藏的LED照明系统，使得天窗无论在白天黑夜都看起来一样精彩。

另外，在空间设计里融入了一些活跃有趣的因素，例如更衣室和咖啡厅，设计方就将它们做成了“封闭的盒子”。这种设计使室内空间看起来更加开阔，协调了与大厅及会议室之间的关系，并采用一层隔音软膜吸收和减弱了来自大厅的噪音，使里面的会议室更显安静和肃穆，完成了从一个开放的都市空间到一个私密的封闭室内空间的过渡。由此拉开第三序列的帷幕。会议室的场景设计也富有戏剧性，像一个方正的盒子，采用轻质木材的地板、墙壁和天花板，营造了一个更合理、隐秘和舒适的环境。单一材料的特性加强了空间的特殊比例，使其墙壁平仄有趣的特点更加显著，当这种精致的纹理和几何图形在木材表面产生光影效果时，整个空间顿时严肃高雅，又奇幻美丽。

This is a transformation of design, design a comprehensive manner taking into account three spatial sequence associated with the transition and integration, and the overall interior space concise compact, but also no shortage of lively and flexible design. Lobby for public space outside its entrance a sign clearly visible, often a transparent display window has been replaced by a large number of black mirror glass, reflecting the city hustle and bustle of the street. From the hallway into the interior space, large red mirror the enthusiasm of the glass timber was warm and light package, all kinds of interesting episodes in this staged a variety of highly expressive values in this integration. The design is also unique features, skylights, full use of natural light, using four real natural light, with the four artificial light source, hidden inside the ceiling LED lighting systems, making roof looks the same in both day and night are wonderful.

In addition, the design inside the space into a number of interesting dynamic factors, such as changing rooms and cafes, designer will be that they made a "closed box." This design makes indoor space seem more open, coordinated lobby and conference rooms with the relationship between and using a layer of soft-film absorption and reduced noise from the hall, the noise, so that the conference room inside even more quiet and solemn completed the city from an open space to a private enclosed interior space transition. Thus the curtain opened the third sequence. Conference room design and dramatic scenes, such as a Founder of the box, using light wood floors, walls and ceilings, creating a more rational, secretive and comfortable environment. Single-space properties of the materials to enhance the specific proportion of its more significant features of the walls, when such a delicate texture and geometric shapes in the wood surface of light and shadow effect, the entire space at once serious and elegant, but also a beautiful fantasy.

↑公共过道的背景装饰墙 / The background of the public hallway decorated walls.

↑连接多个功能空间的公共区域 / To connect multiple functional space in public areas.
↓小型接待室内部 / A small reception inside；↓被红色玻璃围合起来的小型接待室 / Was red glass Wai together a small reception room.

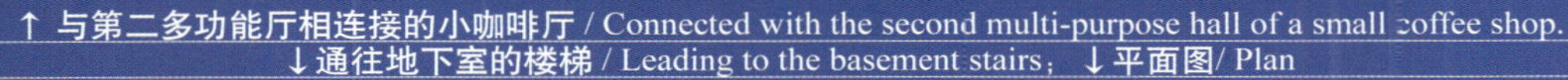

↑ 与第二多功能厅相连接的小咖啡厅 / Connected with the second multi-purpose hall of a small coffee shop.
↓通往地下室的楼梯 / Leading to the basement stairs；↓平面图/ Plan

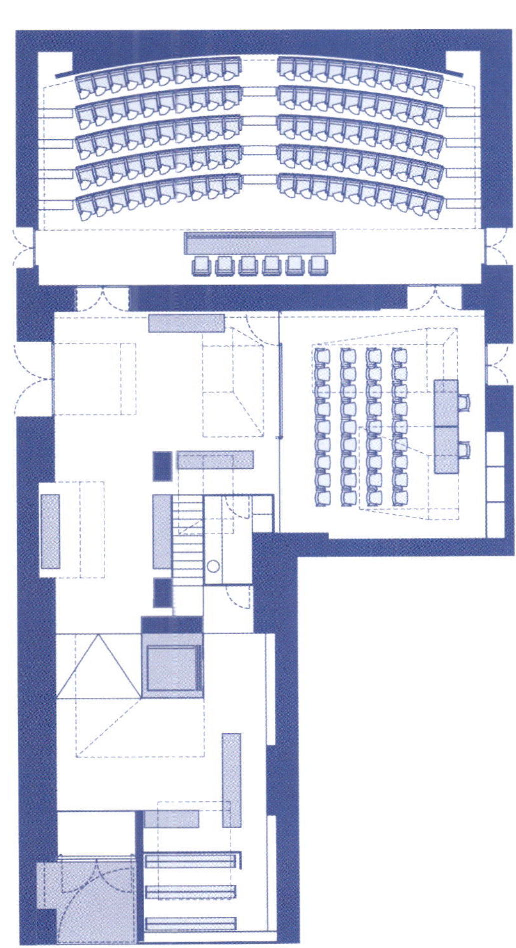

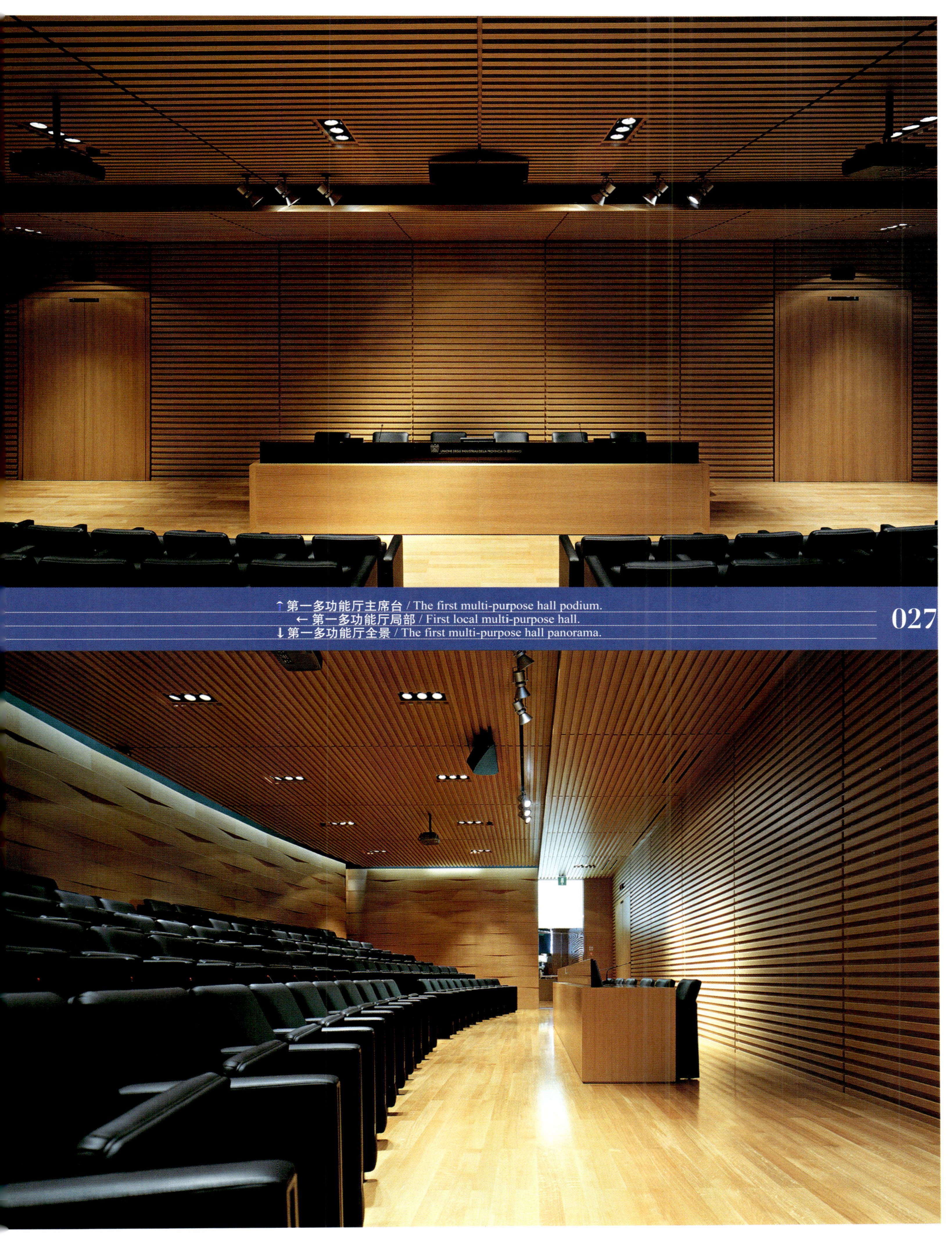

↑第一多功能厅主席台 / The first multi-purpose hall podium.
← 第一多功能厅局部 / First local multi-purpose hall.
↓第一多功能厅全景 / The first multi-purpose hall panorama.

FORTUNE SECURITIES

华宝证券营业部

04

【工程名称】华宝证券杭州营业部 和上海营业部
【面积】500m² （杭州）；420m² （上海）
【设计】胡汉淞；【设计单位】上海元典建筑设计咨询有限公司
【摄影】王瑞璠

本案的设计构思源自于业主对服务精神的认知：率真、直接与真诚的服务。所以在本案的设计中，舍去奢华的视觉效果，以现代主义的空间美学为基础，企图塑造出一个清晰、简洁、穿透的现代空间。

杭州营业所的空间是一个约500m²的长方形平面。面积虽不大，但各种机能都十分齐全。入口的左边是交易厅，设有贵宾交易室、网上查询区、活动交流区等空间。投资者在这个安静典雅的环境里，可以轻松的理财。入口的右边是一休憩区，设有自助吧台，方便投资者在此闲聊、小憩与互动交流。而在接待台的后方，有一间约25 m²的会议室，是客服人员与投资者脑力震荡的场所。所有在这里的活动——会议、交流、研讨乃至于茶会，都可透过两扇活动门，从会议室延伸到其他功能区域，是一个设想周密的弹性空间。

上海营业部的空间约420m²，功能以私人贵宾证券交易室为主。另设有接待处及员工办公区。这两种公私性质不同的区域以一堵中间段被"虚化"的主墙体加以界定，两边各自保留着自己的空间特质及功能，但在视线上又可以穿透流动。这堵中段被虚化的"界定墙"，上半段的功能是照明灯具，而下半段则是艺术品的展示台，各自发挥着被赋予的功能，同时又相互补充。

Case design concept derived from the spirit of the owners of service awareness: straightforward, direct and sincere service. Therefore, in this case design, rounding extravagant visual effects to modernist aesthetics of space, based on an attempt to create a clear, concise, and the penetration of the modern space.

Business office space is a rectangle of about 500 m² flat. Area is not large, but a variety of functions are very complete. The transaction to the left of the entrance hall, equipped with VIP trading room, on-line query area, the activities of the exchange zones of space. Investors in the quiet elegance of the environment, can easily finance. The right side of the entrance is a sitting area, equipped with self-service bar, to facilitate investors in this chat, rest and interact. Reception Desk in the rear, there is one of about 25 m² of conference rooms, is customer service personnel and investors in place mental shock. All activities here - meetings, exchanges, seminars and even tea, can, through the activities of two doors from the meeting room extension to other functional areas, is a well-conceived flexible space.

Sales department of space about 420 m², features a private VIP room-based securities trading. There are also a reception area and staff office. Both public and private nature of the different regions in the middle wall segment was "nominal" to define the main wall on both sides, each retains its own characteristics and functions of the space, but can also penetrate the flow on the line of sight. This is blocking the middle of being virtual-oriented "is defined in the wall", the first half of the function of lighting, while the second half of the segment is the art of the Showcase, each play is given the function, at the same time complement each other.

↑接待前厅 / Reception hall.

↑↓杭州营业部接待台 / Hangzhou Sales Department Front Desk.
↓25 m²的会议室 / 25 m² of conference rooms.

↑休息区 / Rest area.
↓杭州营业部平面图 / Hangzhou Sales Department plan.

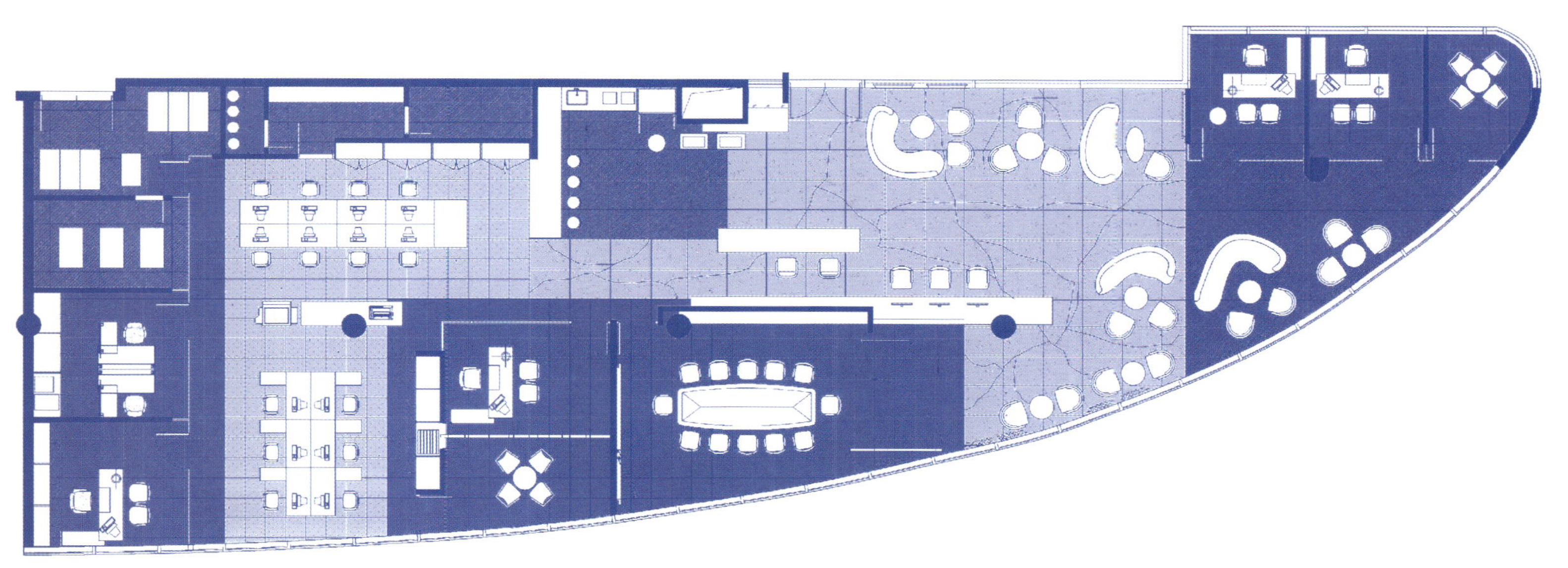

↑上海营业部的接待前厅 / Shanghai Sales Department reception lobby.

↓分界墙 / Boundary wall; ↓从接待前厅看向会议室 / View from the reception hall to the conference room.

↑会议室 / Conference room.
↓上海营业部平面图 / Shanghai Sales Department plan.

SKYHIGH CREATIVE PARTNERS

“天比高创作伙伴”培训中心

05

【坐落地点】香港天水围；【面积】1 042 m^2

【设计】何世梁；【参与人员】Lourance Leung，Kenny Leung，Zeno Yu

【设计公司】何世梁建筑设计公司

【摄影】Alice Ng，Douglas Ho

挑战公认的规范，打破常规的单调，创造出一个思考的盒子，是本项目的设计主题。原本的场地（停车场）建筑像个单调的“盒子”，若要改变表皮的呈现方式，进而形成若干个不同的功能区域，至关重要的就是突破这个“盒子”。

在这个项目中，采光问题总是被优先考虑，外围的墙体只要与房间相连接，设计师尽可能在其上挖出窗户，以便中心区域在日间也能够很好地采光。同样也是考虑到自然采光，主要功能空间都采用开放式，同时其独立性也得到很好的保证。不同于一般的办公室，培训中心设置了如直播室、录音棚、高质量乐队室和控制室等一系列轻松的空间，这样老师与学生之间可以近距离互动。工作区的隔板由两块树脂玻璃中间加一块不透明的壁脚板制成，灯光开启时，工作区好像“浮”在地面之上，轻盈而富有活力。

中心空间层高低矮，因此天花部分的处理在整个设计中扮演了重要的角色。六边形模块天花无形中定义了中心工作区，并为这个大空间提供充足的照明。两排看似浮动的储物柜既统一又各自独立，储物柜由两种颜色的塑料制作而成，“悬浮”在空中，面朝不同的方向。储物柜除了放置物品，还作为一种特殊的隔断，将穿行于该空间中的人流分离，使工作区免受干扰。

Challenge accepted norms, to break the monotony of routine, creating a reflection of the box, is the design theme of this project. The original site of the building like a monotonous, "box", to change the presentation of the epidermis, thus forming a number of different functional areas is essential to break this "box."

In this project, the lighting is always a priority issue, as long as the outer wall is connected with the room, designers dug up the windows on it as far as possible in order to central area during the day can also be a good light. Taking into account the natural lighting, the main function of an open space, while its independence has also been a very good guarantee. Unlike most offices, training centers set up such as live rooms, recording studios, high-quality band room and control room, and a series of easy space, so that could be close between the teacher and student interaction. Work area partitions from the two middle of an opaque Plexiglas wall made of feet, lights turned on, work area like a "floating" above the ground, light and dynamic.

Center space story is relatively low, so part of the treatment of smallpox throughout the design plays a very important role. Hexagonal module smallpox virtually defines the central work area, and for the large space to provide adequate lighting. Two rows of lockers appear to float both the unity they are independent of each of two colors of plastic lockers made of, "floating" in the air, facing in different directions. In addition to lockers to place materials, as well as a special partition, will walk through the flow of people in the space of separation, so that interference from the work area.

↑接待台，倾斜的体块改善了空间的单调性 / Front Desk, tilt body mass improved space monotonicity.

↑入口标志十分显眼 / Entrance signs are very prominent.
↓中心工作区，六边形的座位隔断十分轻盈 / Director of Office of the hand drawing.

↑ 六角模块天花，照明装饰一举两得 / Hexagonal modules smallpox, with lighting and decorative role.
↓平面图 / Plan

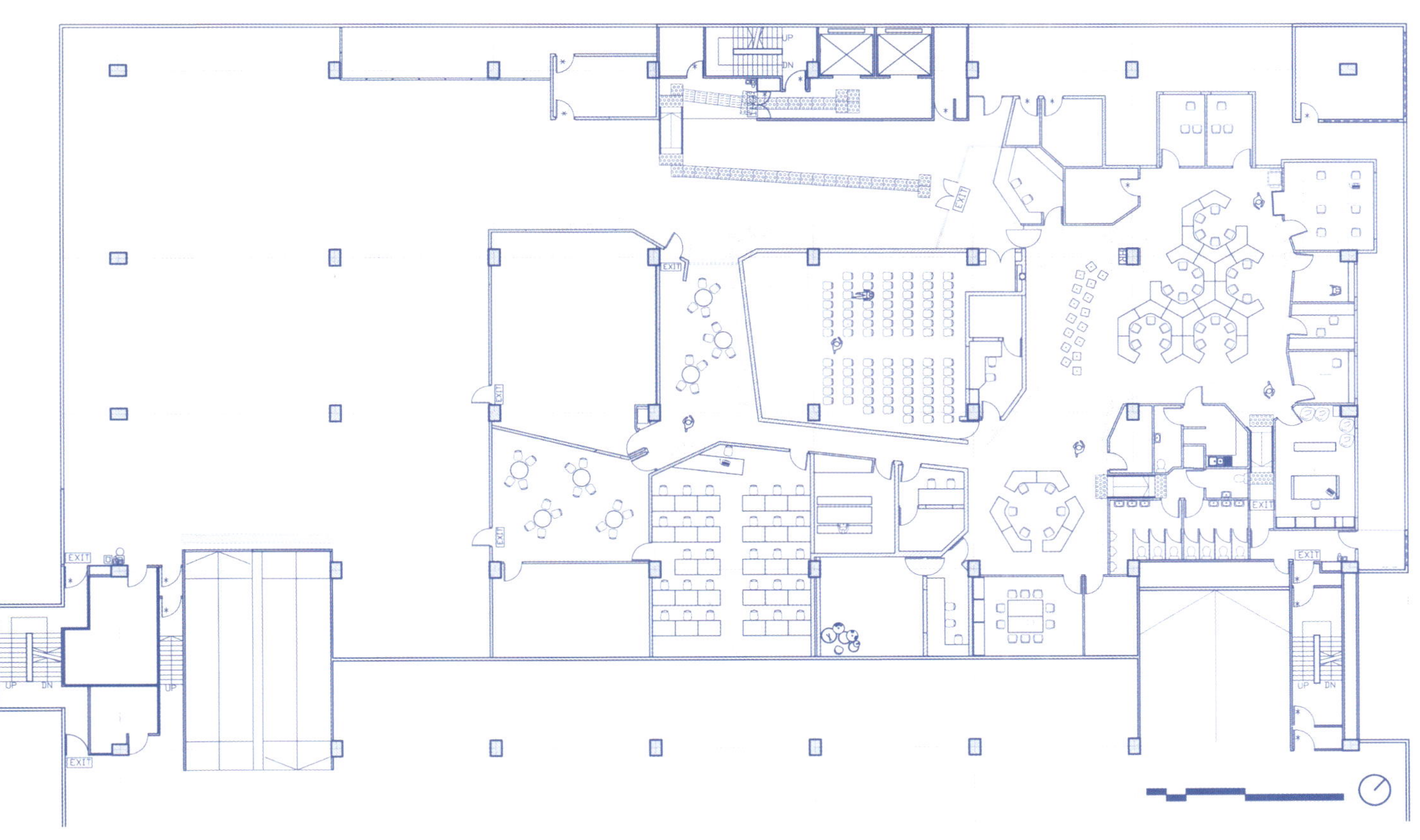

↑头脑风暴区，可进行非正式的会议 / Brainstorming and can be used for informal meetings；↑走廊 / Corridor.
↓倾斜的墙面 / Sloping wall.

↑中心工作区旁边是主管办公室 / Central work area next to the head office.
↓色彩鲜艳的储物柜 / Colorful lockers; ↓宽敞的空间可供运动游戏之用 / Spacious space for sports game purposes.

MERCHANTS MARITIME

深圳招商海运中心

06

【坐落地点】广东省深圳市南山区
【面积】56 500 m²
【设计】刘红蕾、李晓红、randall
【摄影】方振鹏、钱翔

口岸楼大堂是个狭长的4层高空间。平面成一个干净的矩形，空间中有几条连桥贯穿。利用层高的优势，设计师在吊顶做了很多盒体，平面呈矩阵的布置方式，但在高低、色彩上富于变化，形成一个丰富的表皮肌理。两侧高耸的墙面，设计师则把它想像成巨大的集装箱，利用错位的条纹，组织成一个丰富的集装箱表皮……

报关大厅的顶部，设计师利用条状彩色铝板的特点，形成了平面化的图案。在组织色彩的时候，米色、白色、棕色为办公室空间注入了温暖的情绪。

主塔楼的大堂为双层高空间，对于中间的核心筒部分，设计师利用表皮化的方块语言雕塑出形体，顶部采用了"黄金分割"，做出一个大小白色方块的构成变化。

Port floor lobby is a long, narrow four-storey space. Plane into a clean rectangular space there are several bridges throughout the company. The use of the advantages of a high level, designers have done a lot in the ceiling box body, flat layout was the matrix, but in high or low, rich color on the changes in the skin to form a rich texture. Both sides of the towering walls, designers put the Think of it as a huge container, using dislocation stripes, organized into a rich container skin ... ...

Customs hall at the top, designers use the characteristics of aluminum strip color, forming a planar pattern. When the organizational color, beige, white, brown for the office space into a warm mood.

The main tower lobby for the double-high space, for the middle part of the core tube, designers using skin-based language sculpture out of the box body, at the top with the "Golden Section", to make a change in the composition of the size of the white box.

↑接待台，倾斜的体块改善了空间的单调性 / Front Desk, tilt body mass improved space monotonicity.

大
厅

↑ 报关大厅的天花设计 / Customs hall, the ceiling design.
← 铁制连桥 / Even iron bridge.
↓平面图/ Plan

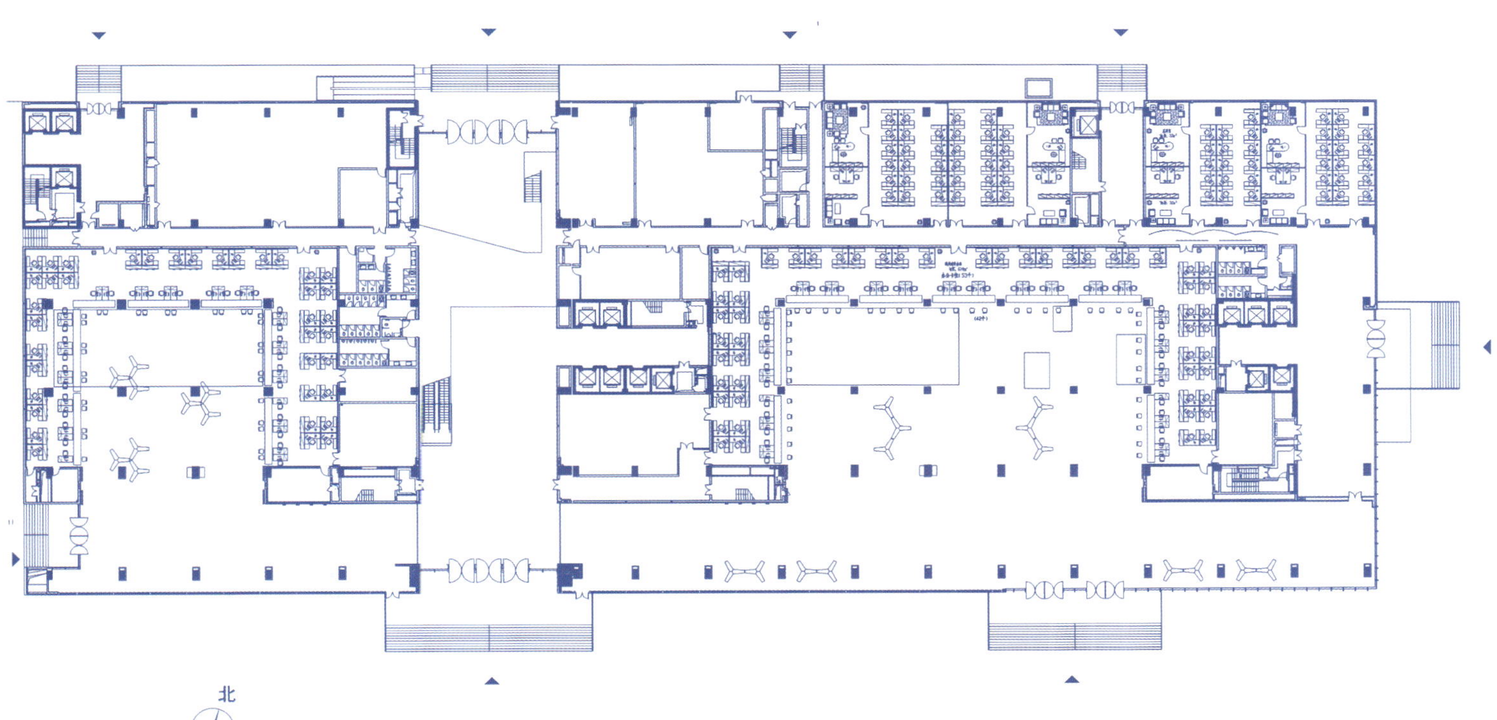

↑报关大厅墙面设计 / Customs hall wall design；↑走廊 / Corridor
←塔楼内部的设计 / Within the tower design.
↓顶棚图 / Roof plan.

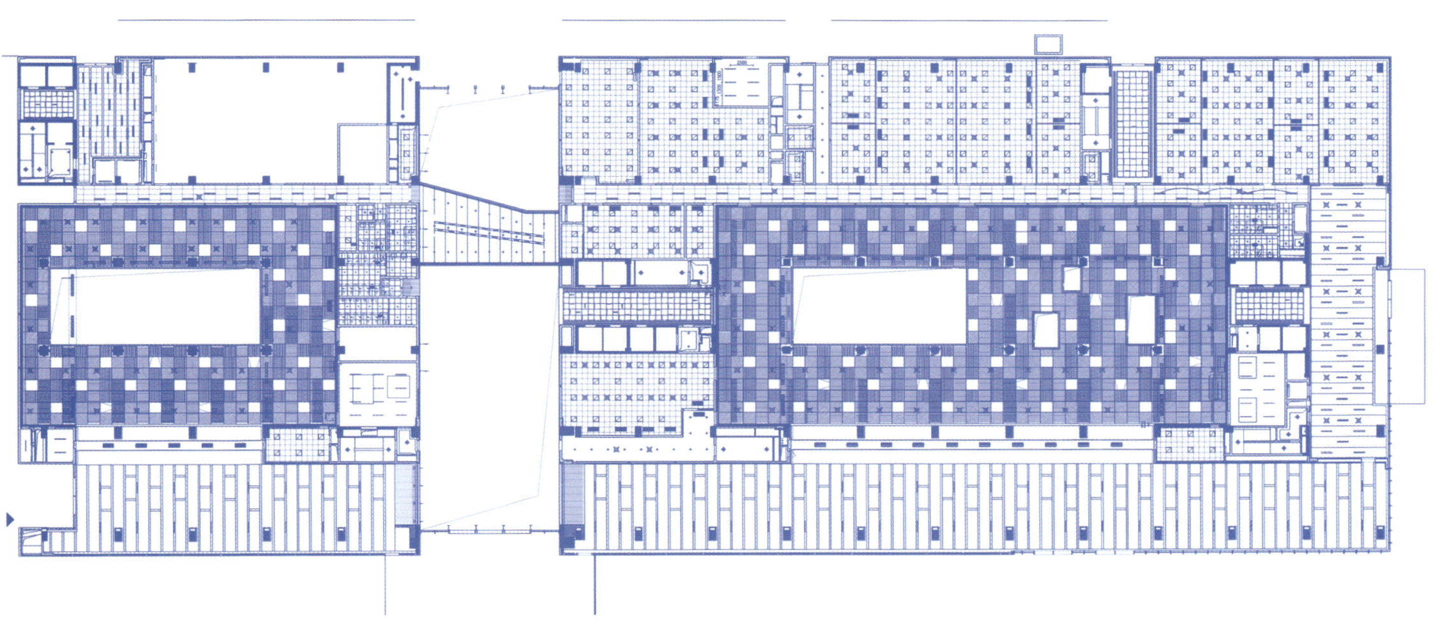

SOHU GAME CHANEL OFFICE

搜狐游戏中国总部

07

【坐落地点】北京石景山

【面积】5 000 m^2

【设计】陈宪淳、生兵

【摄影】贾方、陈宪淳

在整个项目的平面处理上，设计师用到了大量的弧线。这些线段贯穿着这个空间的休闲区域，让密密麻麻的办公空间里面，多了些许呼吸的空间，灵动而有趣。绿色植物的运用也很用心。真正的悬挂式植物墙，让绿化不再停留在摆设的层面，而"滴灌式"系统让植物的护理变得简便而可行。

设计师将这个案例分为两大部分：办公区域及公共服务区域。办公区域包括独立办公室、开放办公区以及机房、储存间等偏重功能性的辅助设施。而公共服务区域则包括前台、接待区、茶水间、休闲区、健身活动区、展示区、会议区等偏重娱乐与接待功能的区域。这样前者可以尽可能地利用原有的图纸，只做优化调整及配色处理，而将主要的精力放在后者上。

针对搜狐游戏这个企业本身的年轻活力及丰富创造力的特点，设计师营造出一个充满激情与浪漫的空间。这也同时可以与相对保守的传统开放式办公区现成极大的反差，有利于减弱员工工作的烦闷情绪从而提高工作效率。由色彩跳跃的夸张雕塑造型来作为空间的主导元素，同时也结合室外植物的培育及展示手法来突出健康与环保主题，是这个设计最突出的特色之一。

Plane handled throughout the project, the designers used a large number of arc. The segment runs through the space of leisure areas, so that inside the crowded office space, an increase of breathing space, flexible and interesting. The use of green plants is also very carefully. The real hanging plant wall, so that green is no longer remain at the level of decoration, and the "drip-style" system allows the care of plants has become simple and feasible.

Designers this case is divided into two parts: the regional office and public service area. Office area includes an independent office, open office area, as well as rooms, storage rooms, and so emphasis on functional ancillary facilities. The public service areas include the front desk, reception area, tea rooms, recreation area, fitness activities area, display area, conference area features such as emphasis on entertainment and reception area. The former can use this as much as possible the original drawing only did optimal adjustment and color processing, and the focus will be mainly on the latter.

Sohu games for the enterprise's own youthful vigor and rich characteristics of creativity, designers create a room full of passion and romance. This is also relatively conservative with the traditional open-office ready a great contrast, there is less conducive to employee feelings of anguish and thus work more efficiently. Exaggerated by the colors jump sculpture forms as the leading element of space, but also the cultivation of plants combination of outdoor and display techniques to highlight the health and environmental topics, is the most prominent feature of the design.

↑吧台、地台、局部天花组成连贯放的造型 / Bars, slabs, partial release of smallpox form a coherent shape.

↑入口等候区 / The entrance waiting area.
← 翠绿LOGO背景墙 / LOGO green backdrop.
↓贵宾接待室 / VIP reception；↓轻松自然的茶水间 / Easy natural panty.

↑ ↑休闲区、展示区、健身活动区三者的空间关系 / Recreation area, display area, fitness room for the relationship between the three activity areas.

→墙上是立体的小白菊花 / The wall is a three-dimensional small white chrysanthemums.

↓二层平面图 / 2nd floor plan； ↓三层平面图 / 3rd floor plan.

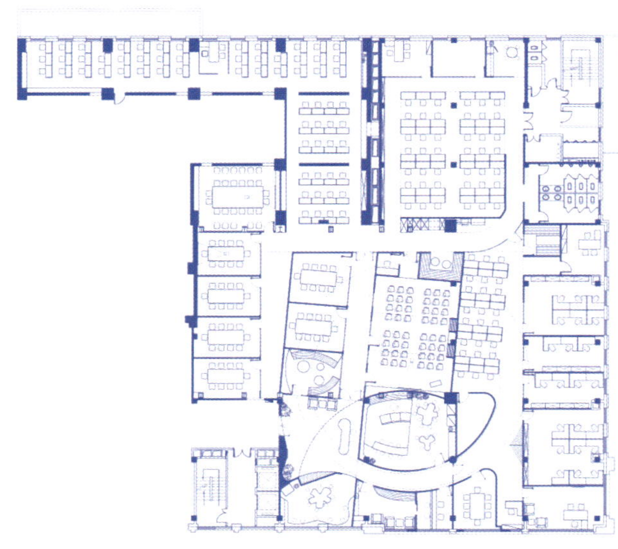

↑吧台剖面 / Bar profile.

←左边是弧形的展示区，右边是会议区的造型墙 / The left is a curved display area, conference area on the right is the shape of the wall.

↓四层平面图 / 4th floor plan；↓五层平面图 / 5th floor plan.

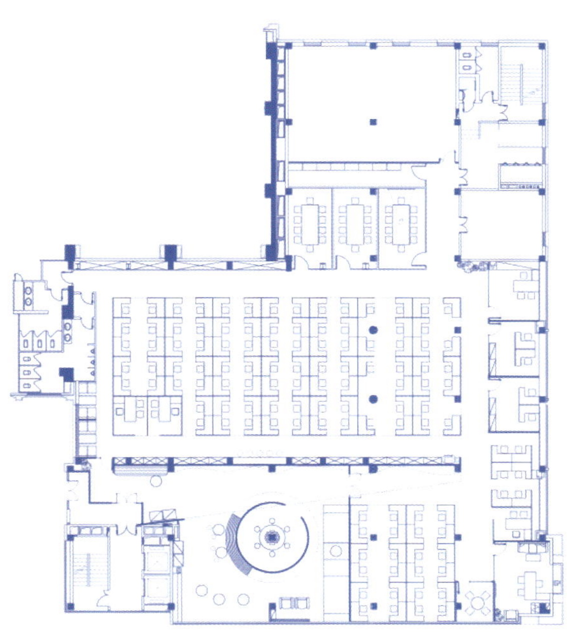

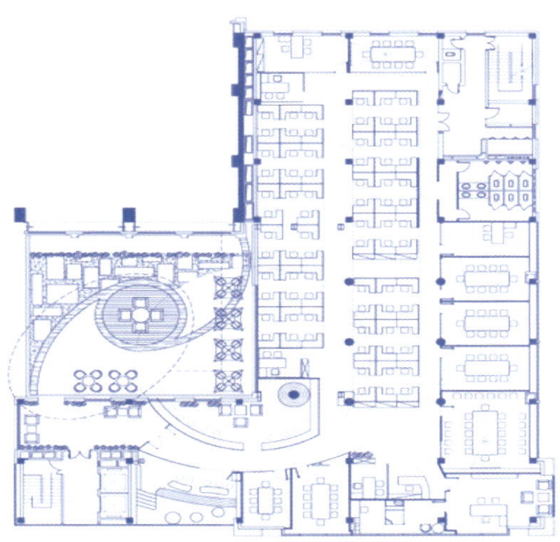

↑无论墙体还是空间设计都别出心裁 / Whether the wall and space designs try to be different.
← 室外植物的培育及展示手法来突出健康与环保主题 / Outdoor cultivation of plants and display techniques to highlight the health and environmental topics.
↓休息区剖面图 / Rest area profiles.

AN DER ALSTER 1

An der Alster 1综合办公楼

08

【坐落地点】德国汉堡；【面积】970 m²
【设计】J. MAYER H. 建筑师事务所〔德〕
【主要材料】混凝土、石膏板、玻璃
【摄影】FotografieSchaulin, Hiepler Brunier, Dirk Fellenberg, Schraubverschluss

建筑的外立面十分独特，大面积采用玻璃，白色的水平线将楼层划分出来。其中几块凸出的部分打破了本来的平衡。办公楼的前面是一个精心设计的广场，绿草坪上穿插着曲折的小径，椭圆形的小花坛点缀其中，与建筑的外立面有异曲同工之妙，二者和谐地融为一体。

室内空间中包含两种平面布局，一种是常规的设计，就如同一般常见的办公空间一样；另外一种则是在“眼睛”的位置进行特殊的设计。大跨度的空间与阳台结合，为设计师的创作提供了多种可能性，“眼睛”的内部是一个相对较大的空间，存在于建筑双层表皮的后面，室外的冷暖空气进入室内的时候在这里得以缓冲，从而调节了室内气候的平衡。

这种表皮设计方式能够有效地节约能源，“眼睛”——也就是内部的窗户向外凸出的部分，其内部可以用来安装通风设备，这样就不需要额外增加空调设备。除此之外，混凝土天花板内部的混凝土芯活化系统在水动力协助下，使建筑内部在夏天更凉爽，冬天更温暖。同时，保暖性的外表皮也可以降低热能耗，这样不但确保了人体的舒适度，而且最大限度地节省了能源。

Building's facade is very unique, large-scale use of glass, a white horizontal line would be separated floors. A few of them broke the protruding part of the original balance. The front office is a well-designed square, green lawn interspersed with winding paths, a small oval-shaped flower embellishment which the same role with the structure's facade, the two blend harmoniously.

Interior space contains two kinds of layout, a conventional design, just as in general the same as a common office space; and another type is the "eyes" of the location of special design. Large-span space and balcony combination of creative designers offer a variety of possibilities, "eyes" of the house is a relatively large space, exists in the building behind the double-skin, outdoor heating and cooling air into the room when the be a buffer, which regulates the indoor climate balance.

This skin design approach can effectively save energy, "eyes" - that is, outside the windows inside the protruding part of the internal can be used to install ventilation equipment, so that no additional air-conditioning. In addition, the concrete ceiling within the concrete core activation system in the water-powered help, so that the internal building cooler in the summer, warmer in winter. Meanwhile, the warmth of the outer skin can also reduce the thermal energy, so as to ensure the comfort of the human body, but also to maximize energy savings.

↑建筑外立面 / Facade

↑高达两层的入口大门 / Up to two-storey entrance gate.
←纯白的会议室 / Pure white conference room.
↓一层平面图 / Floor plan.

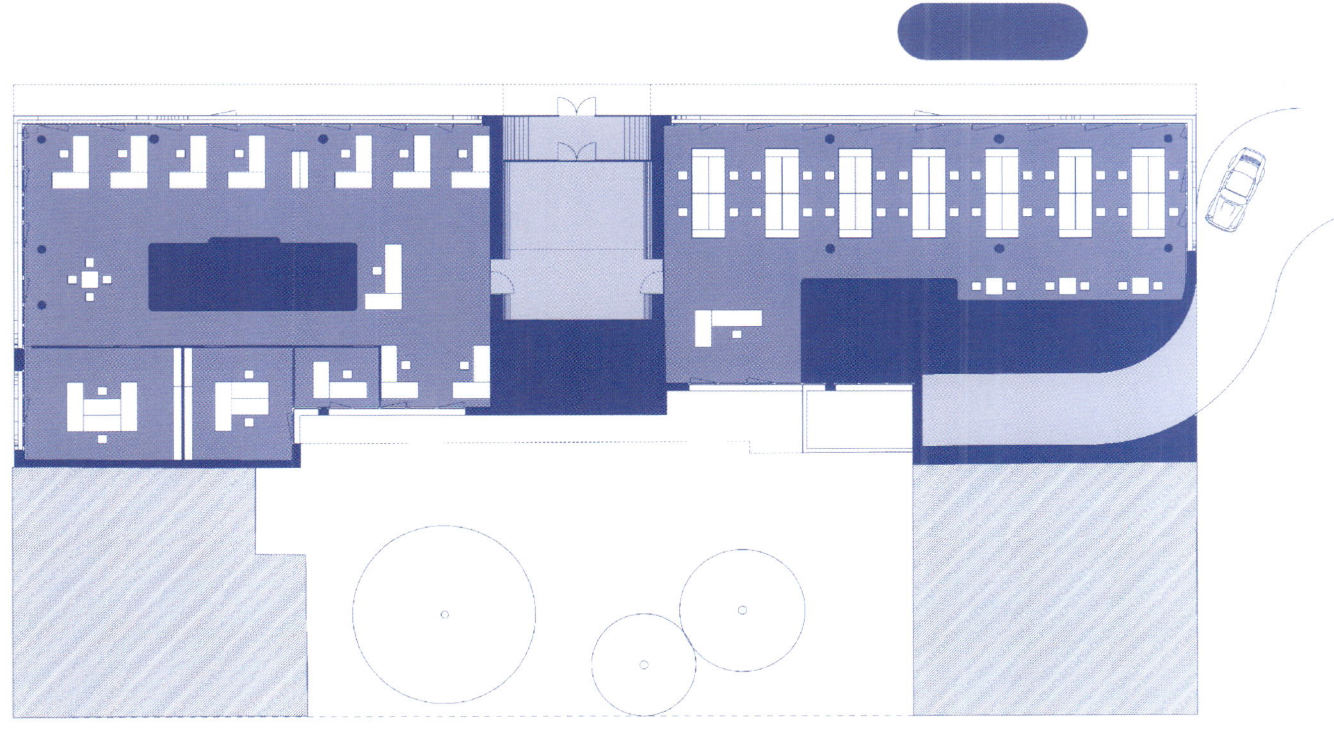

↑玻璃隔断中的会议室更显玲珑剔透 / The glass partition in the conference room is even more exquisitely carved.

→会客室一角，天花与凳子的设计上下呼应 / Reception room corner, ceiling echoed up and down with the bench design.

↓二层平面图 / 2nd floor plan.

↑大面积使用落地玻璃窗，室内光线充足 / Large-scale use of floor windows in sunlight.
← 楼梯强烈光线的照射下变得十分梦幻 / Exposure to intense light under the stairs become very dream.
↓三层平面图 / 3rd floor plan.

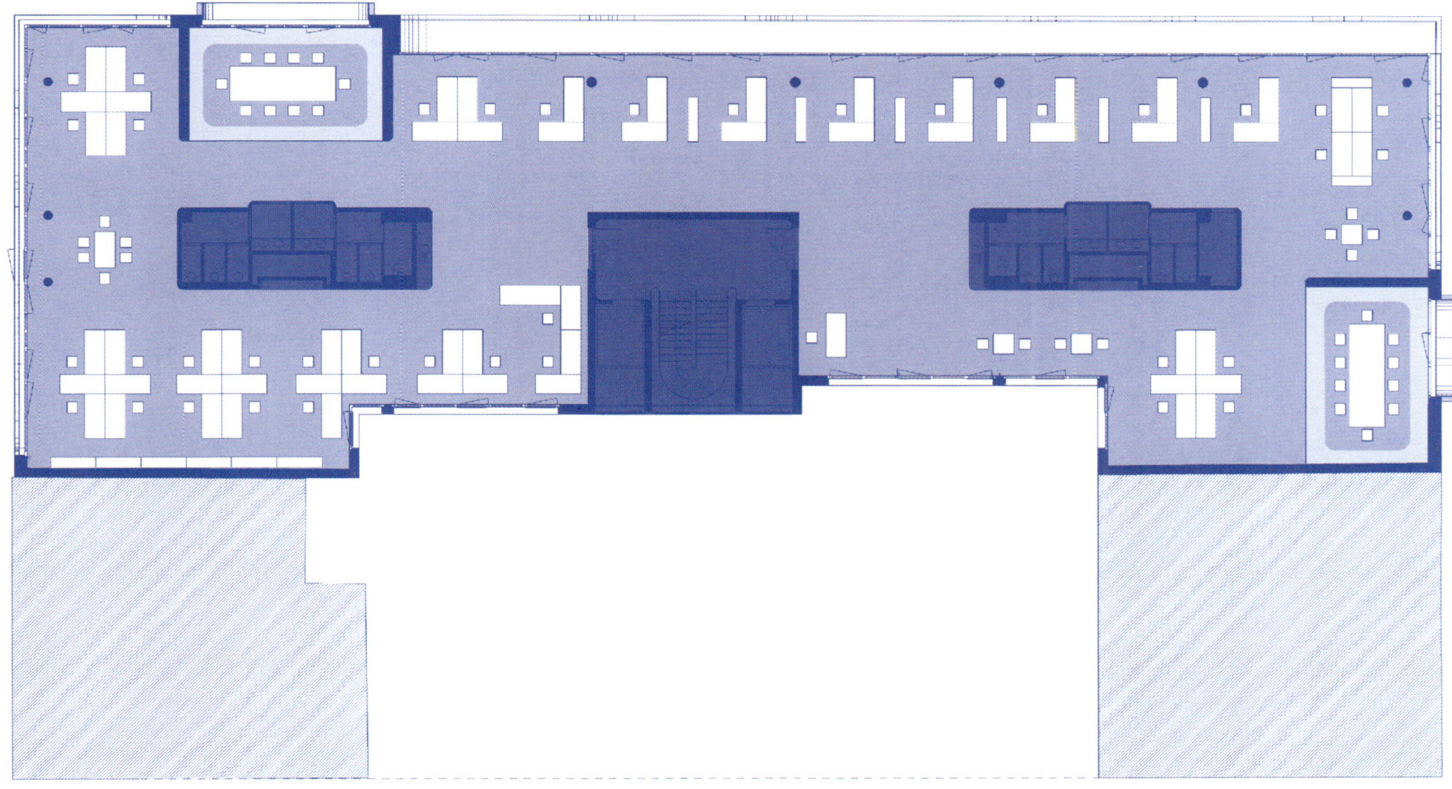

BLACKWELL SCIENCE ASIA OFFICE

布莱克维尔科学亚洲公司

09

【坐落地点】澳大利亚墨尔本；【面积】2 200 m^2
【设计】澳大利亚Williams Boag建筑设计事务所
【项目人员】Catherine Ramsay, Kerry Smith, Peter Williams
【摄影】Tony Miller

方案着眼于将现有的空间改造成开放、明亮的室内环境，内部有一个中心区域，员工与客户可以在此聚集。呈几何圆形的中庭空间是建筑方案的重点，它被打造成整个办公楼的枢纽地带，员工与来访者可以在这里会面、参观、讨论公司的产品。空间的另一个关键元素是出版物陈列及有图解说明的背光展示柜。

由临街入口到前台，然后穿越中庭至公司各部，地面是橡胶材料，墙面是金属板和塑胶玻璃板，它们构成一种平面线性组合以及条带状的视觉效果。所有的调整与细部处理都为突出中庭和强化圆形枢纽的作用。毗邻前台及中庭的会议室和员工区域均由活动隔音板分开，在进行全体员工会议或大型推广活动时，可以在近邻中庭的位置迅速安排出一个很大的空间来。在会议室及员工区域选用可活动的家具，最大程度上强化空间的多功能性，带脚轮的桌子根据需要可以自由组合，或折叠推至一旁。

墙面一般采用米黄色调涂料以增强照明反射度。中心色调主要体现在储存柜黄色漆面板、办公区屏风、候客厅家具面料和定做的地毯这几个大面积的色块上。地板材料一般使用橡胶、软木板和地毯砖，以平行的带状铺设来指示公共区、办公区、配套区。

Program is aimed at transforming the existing space into an open, bright indoor environment, the inside has a central area, employees and customers can be gathered here. Geometry was circular atrium space is the focus of the construction program, which was created to fight the hub of the entire office area, staff and visitors can meet here, visits to discuss the company's products. Another key element of the space exhibition and publication are the backlight display cases illustrate.

From the street entrance to the front, and then across the courtyard to the company's departments, the ground is a rubber material, the wall is a metal plate and plastic glass, which constitute a linear combination of planar and strip the visual effects. All the adjustment and detail treatment is to highlight the circular atrium and strengthen the role of a hub. Adjacent to the front and in the atrium of the conference rooms and staff activities in the region barriers separated by conducting staff meetings or large-scale promotional activities, you can place in the neighboring courtyard quickly arrange a lot of space. In the conference room and staff activities of the regional selection of furniture may be the greatest degree of versatility to strengthen the space with a table castors are free to mix as needed, or pushed to the side of the fold.

Commonly used cream-colored wall paint colors to enhance the degree of light reflection. Center color is mainly reflected in the yellow paint storage cabinet panels, office wall, waiting room furniture, fabrics and custom-made carpet these large blocks of color on. General use of rubber flooring, soft wood and carpet tiles, laid in parallel strips to indicate the public area, office area, and the supporting areas.

↑主楼梯 / The main staircase.

↑室内大堂 / Indoor hall.
← 主楼梯的几何面运用了水平元素 / The main staircase of the geometric surface elements used to a new level.
↓一层平面图 / Floor plan.

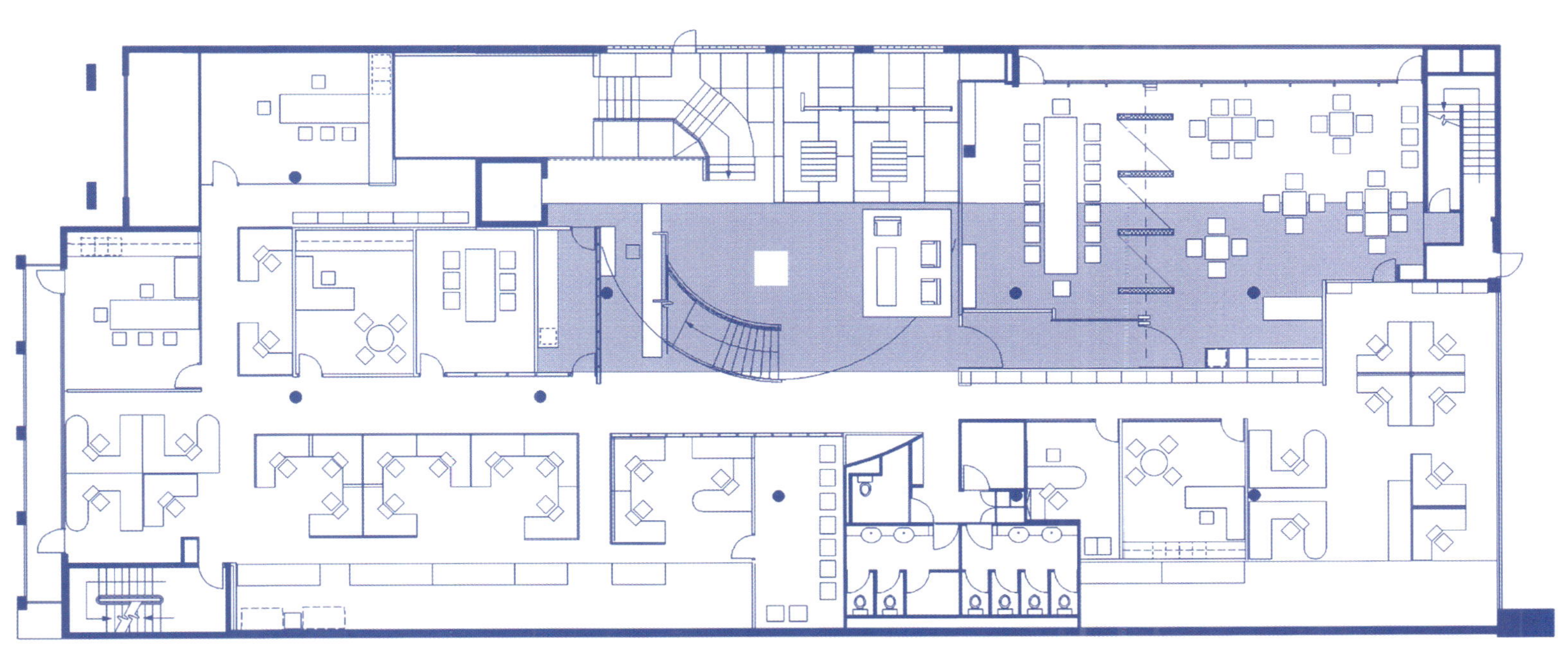

↑空间全貌 / Space picture.
↓剖面图 / Section

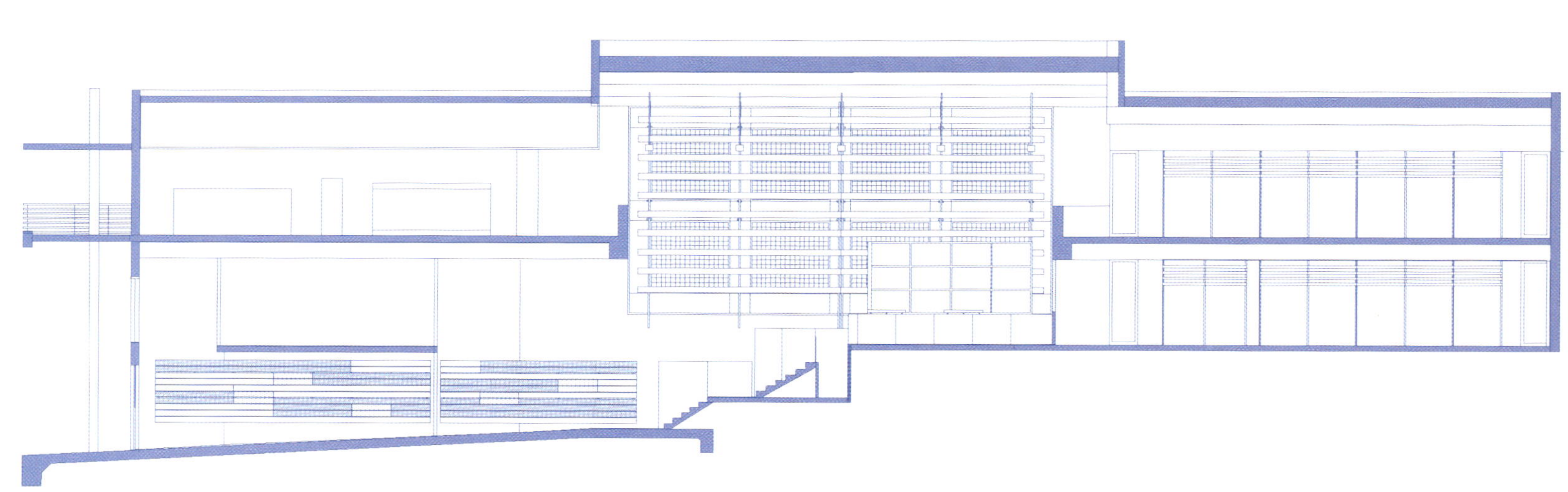

↑空间全貌 / Space picture.
↓二层平面图 / 2nd floor plan.

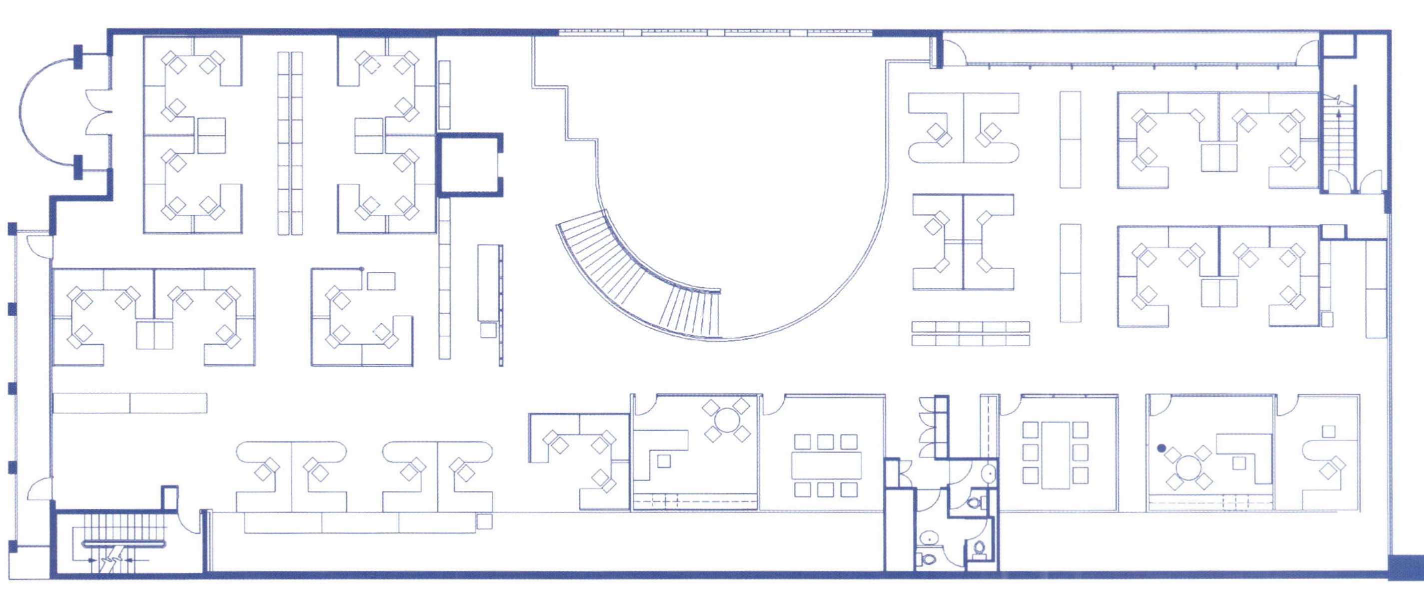

↑布置灵活性的小会议室 / The small room layout flexibility.
← 带会议桌的负责人办公室 / Conference table with a person in charge of office.
↓屏风 / Screen；↓强烈的线性图案 / A strong linear pattern.

ZHENGDA CUBE EDIFICE

证大立方大厦办公室

10

【坐落地点】上海市浦东长柳路58号；**【面积】**A型，185 m²；D型，282 m²
【设计】李玮珉
【主要用材】A型，芝麻灰花岗岩（烧面）、白膜胶合玻璃、钢化清玻璃、上贴喷片、拉丝面不锈钢、地毯；D型，黑色乳胶漆、亮面不锈钢、灰色木纹石、喜马拉雅珍珠黑大理石、地毯、壁纸

整栋楼内的套间都以重叠交错的跃层结构组合而成。大楼外立面的凹凸体块勾勒出与跃层空间相映照的造型，建筑看上去就像是一个形式有趣的立方体。建筑设计的理念虽然新鲜，但是内部跃层空间的格局却为功能的合理性分布制造了不小的麻烦。为了向购买者展示跃层空间使用上的巨大潜力，设计师虚拟了两种可能性，即A、D两套样板房型，既延续了建筑设计理念中趣味的构思，又赋予其功能使用上的合理和便利。

A型样板间设想业主是一个产品设计公司。虽然上下两层的面积总和只有185 m²，但是设计师希望它“麻雀虽小，五脏俱全”。于是，一层的功能被设计为向公众展示设计作品的小型展示空间，二层为办公空间。这样，上下两层既相互关联，又因空间上的物理性划分而各自独立，更加贴合了产品设计公司特定的功能需求。贯穿上下的楼梯无疑是空间的重要部分。

D型样板间设想业主为服装设计公司。与A型样板房将展示与办公分离的设计思路相类似，一层模拟的是该品牌服装的专卖店设计，二层为办公区。两层空间通过一座宽大的L形楼梯联结。

The suites have allowed their entire buildings with overlapping staggered structure of combinations of thermocline. The building facade of the convex-concave body mass outlines of space relative to shine with the thermocline shape, the entire building looks like an interesting form of a cube. Architectural design concept is fresh, but the pattern of space within the thermocline is the distribution of functional rationality has created no small trouble. In order to demonstrate thermocline buyers use the enormous potential of space, designers virtual two possibilities, namely, A, D two sets of sample chamber, a mixture of continuity in the architectural design concept of the idea of fun, but also the use of its functions conferred on the reasonable and convenience.

A-model between the idea is a product design company owners. Although the total area of the upper and lower two floors only 185 m², but the designers hope to meet a lot of features. As a result, a layer of functionality is designed to showcase the design work to the public, a small exhibition space, the second floor for office space. In this way, the upper and lower two are interrelated, and also because the physical space into the separate, more fitting of the product design company-specific functional requirements. Through the up and down the stairs is undoubtedly an important part of space.

Between the D-model for the apparel design company owner envisaged. With the A-model room will showcase design ideas with the office of separation is similar to a layer of simulation is the brand clothing store design, the second floor for office. Two-story space by a large L-shaped staircase join.

↑楼梯的另一角度 / Another angle of the stairs.

↑灯槽的设置 / Light tank setup.

↓A型样板间的一层展示空间 / A-model layer of display space.

↑A型样板间的2层办公区 / A second layer of model-based office.
↓A型样板间平面图 / A-model layout.

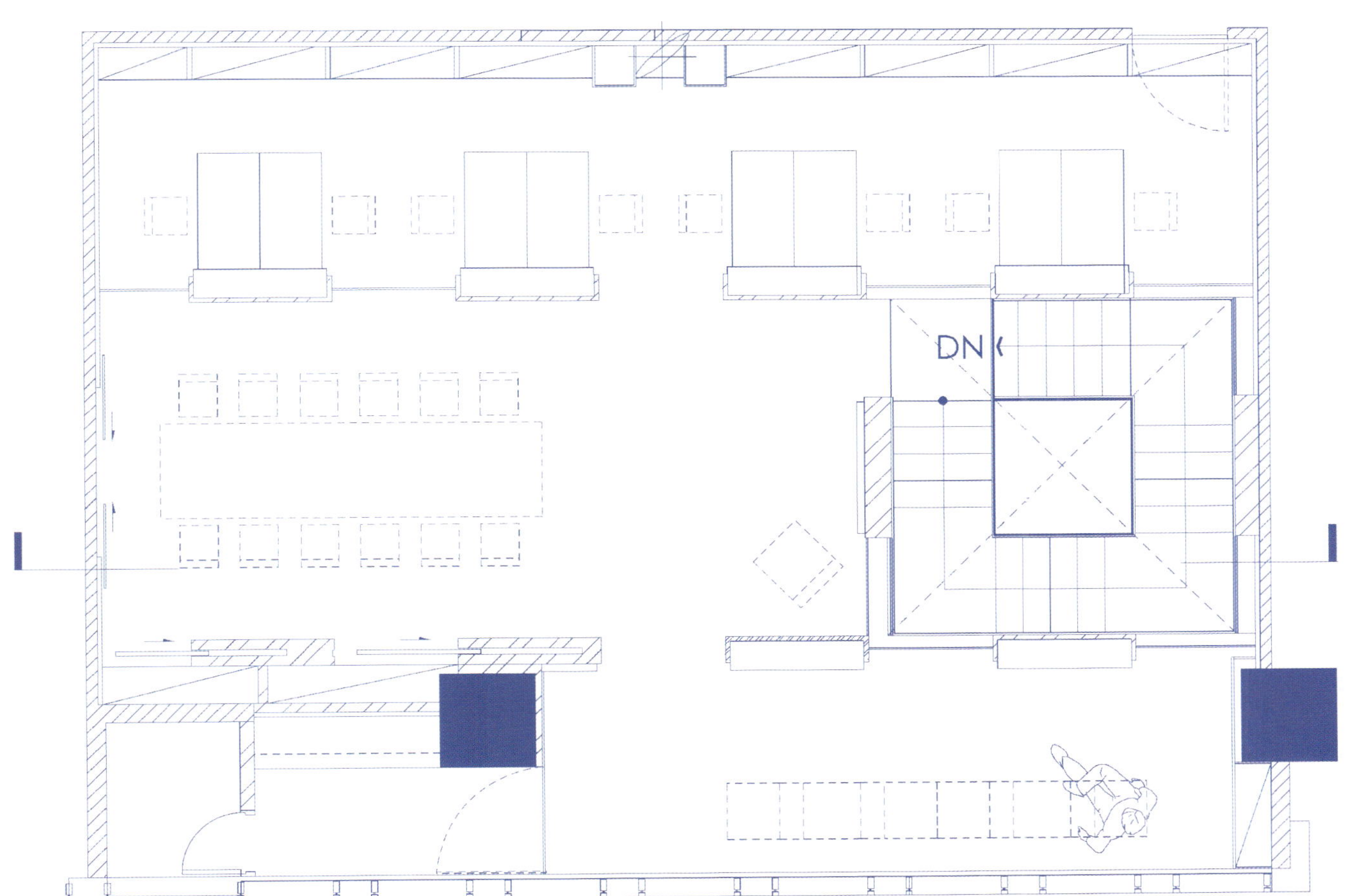

↑A型样板间的2层办公区 / A second layer of model-based office.

↓二层开放的会议空间 / Second floor of an open meeting space.

↑宽大素雅的L形楼梯 / Elegant large L-shaped staircase.
↓D型样板间一层平面 / D-model first floor flat.

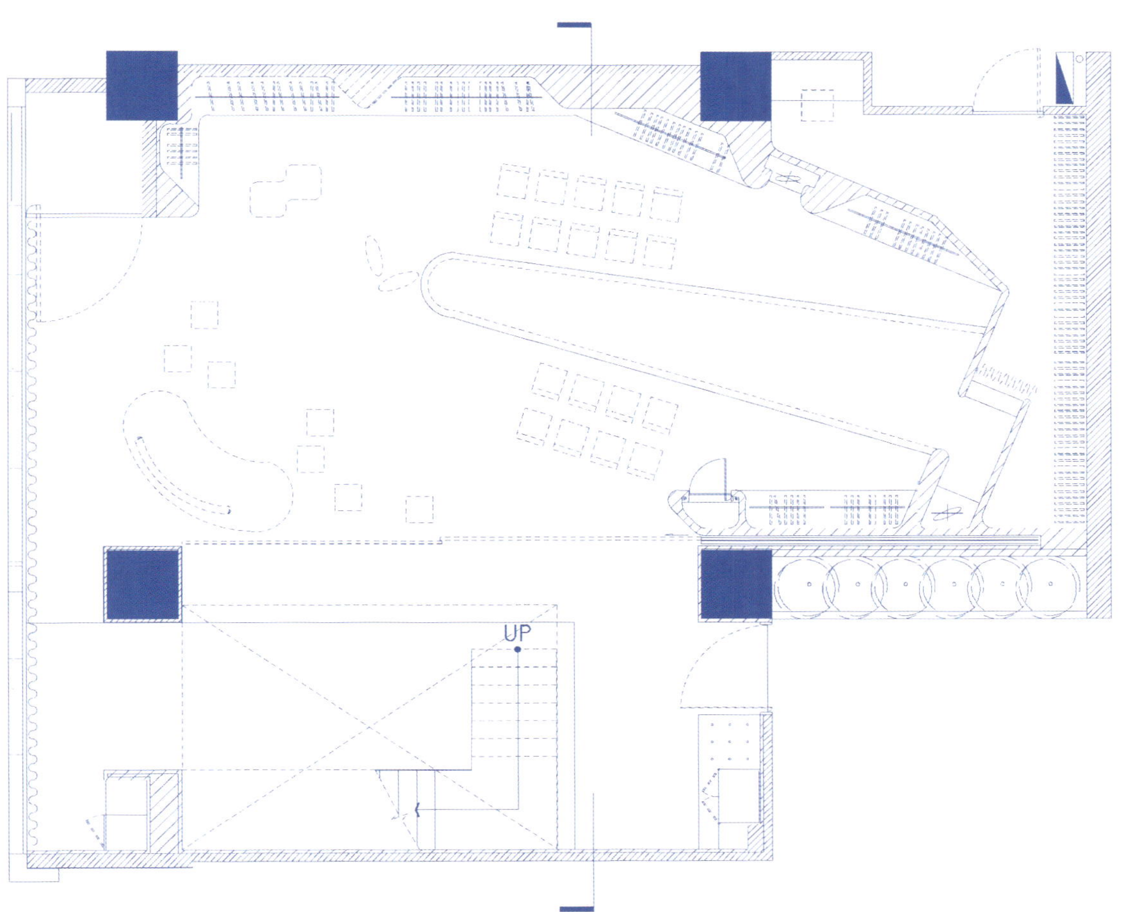

↑D型样板间的一层展示区 / D-model display area.
↓楼梯细部 / Staircase detail.

↑二层淡雅通透的办公空间 / The second transparent layer of elegant office space.
↓D型样板间二层平面 / D-model second floor flat.

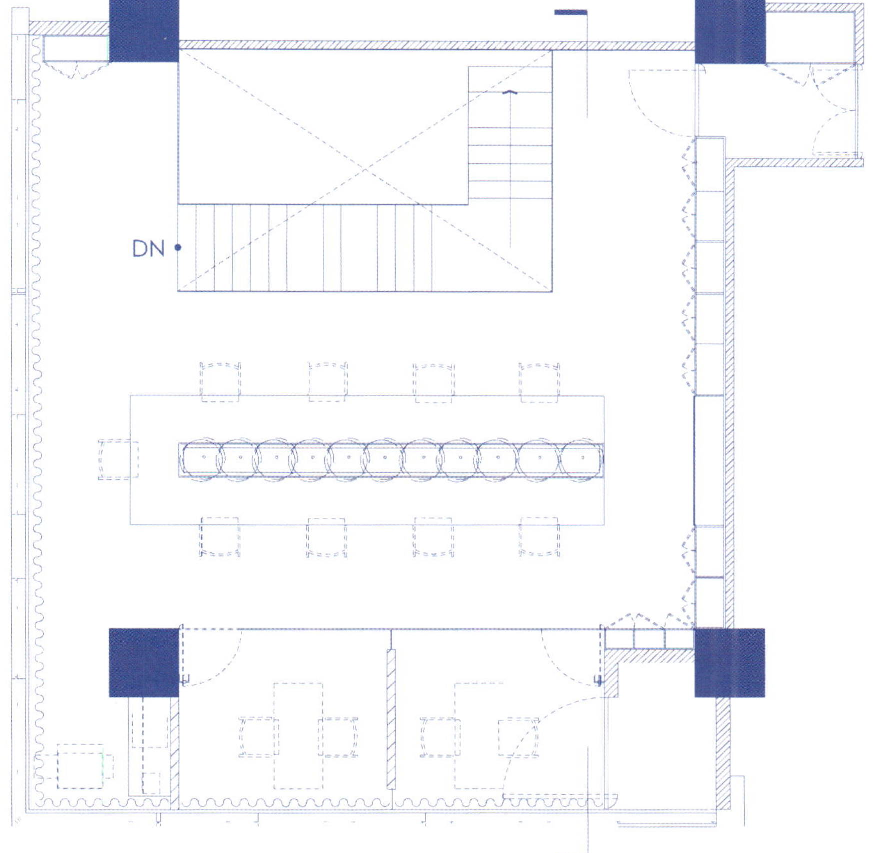

CHINA TIMES MEDIA GROUP

中时媒体集团总部

11

【坐落地点】中国台湾台北市大理街
【面积】8 300 m^2
【设计】林洲民
【设计公司】林洲民建筑师事务所 / 仲观设计顾问有限公司

由于历经50年的分期开发，在办公与作业空间的使用上，各部门分处不同的大楼内，实质空间缺乏整体的串联。因此，设计师设法扩大垂直方向的畅通，进而创造出集团的核心价值：一个可以灵活反应媒体创作迅速变化的开放空间。

设计师将空间作为叙事空间的精神转换，创造视觉的连通、跨部门合作的沟通机制、交流与活动事件的平台、多元信息的载体，最终是一个媒体人的PowerHouse。一个篮球场、半透明的圆形会议室、休闲与交流的阶梯看台，即使不符合空间使用的经济考虑，设计师仍坚持一定比例的公共开放空间，为工作者和整个集团创造更多可能的优质效益。

在这里，人是主角，使用者的创意决定空间的定义；空间是配角，以缪斯之姿，在蜡版上刻记简单的点缀。挑高的楼层设计，将大自然的视野与静谧氛围重现于地底；宽广的编辑平台，串联各部门共谱名为效能的和谐乐章；全新概念的会议空间，在空中、在书海里、在玻璃帷幕内，打破沟通的限制，捕捉源源不绝的精彩论点；遍布的装置艺术，感性诠释过往，以历史为前进未来的篇章脚注；全日开放的运动与公共休闲空间，加倍充电疲惫的身心，运球声和打字声，同样是媒体工作的魅力展现。

After 50 years as a phased development, in the office and work space and the use of various departments within the sub-offices in different buildings, the real lack of space, the whole series. Therefore, designers seek to expand the vertical direction of the flow, thereby creating the Group's core values: a flexible response to rapid changes in media, creative open space.

Designers of space as a narrative space in the spirit of conversion, and create visual connectivity, cross-sectoral cooperation, communication mechanisms, communication platform for events and activities, multi-information carrier, is ultimately a media person's PowerHouse. A basketball court, translucent circular meeting room, leisure and exchange ladder stands, even if space usage does not meet the economic considerations, designers still insist on a certain proportion of public open space for workers and the whole group to create more potential high-quality benefits.

Here, people are main characters, the user's creative definition of the decision space; space is a supporting role, in the paper, engraved in mind a simple embellishment. High-ceilinged floors will be designed to the vision of nature and the quiet atmosphere to reproduce in the underground; broad editorial platform, tandem silver threads, called the performance of various departments and harmonious movement: a new concept of meeting space, in the air, and in the sea, inside the glass curtain, breaking the communication constraints, to capture plenty of exciting arguments; over the work of installation art, sensual interpretation of the past, in order to advance the next chapter in the history as a footnote; full day of sports and public recreational open space, double charging the physical and mental fatigue , dribbling sound and typing sound, is also the work of the charm of the media display.

↑挑高空间 / High-ceilinged space.

M
N

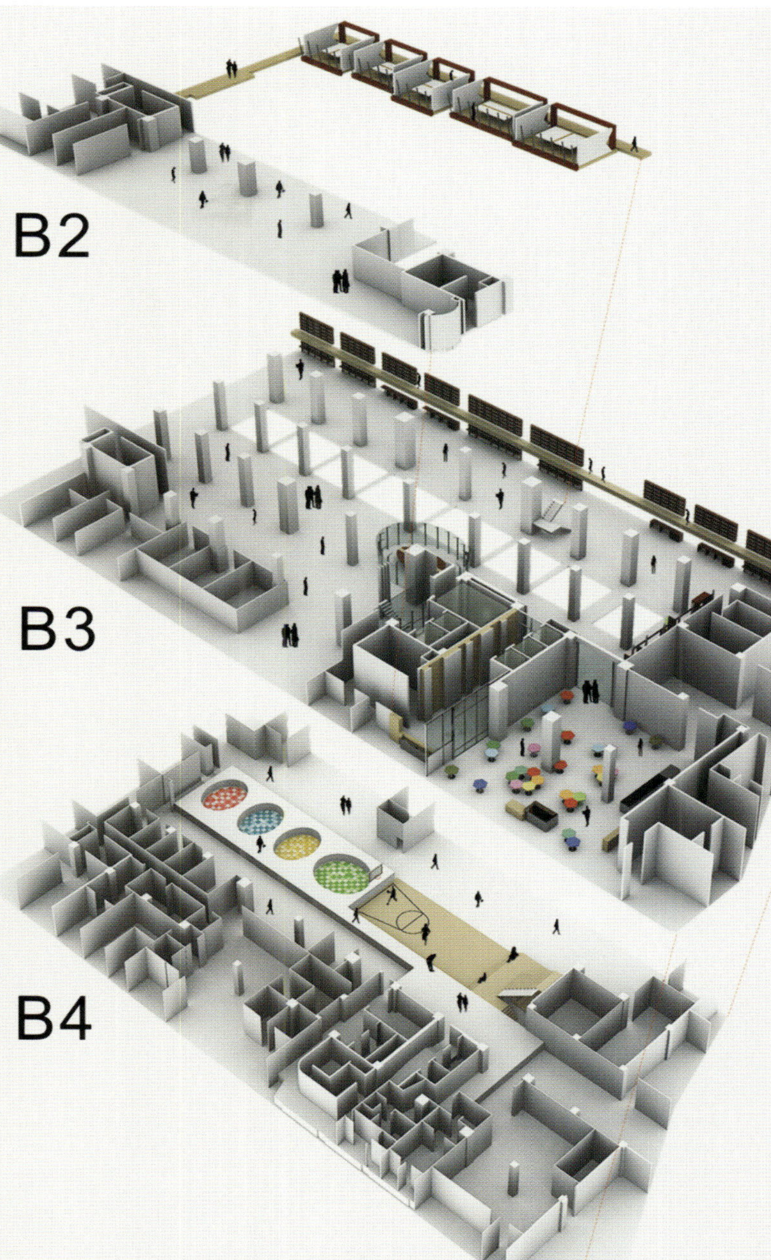

↑设计规划模拟图 / Design planning simulation diagram; ↑B4会议室造型灯 / B4 conference room modeling lamp.
←B3办公空间全貌 / B3 office space picture.

↑B3办公空间 / B3 office space.
↓会议室走廊 / Meeting room corridors.

↑B3书墙 / B3 book wall.
↓会议室 / Conference room.

↑B3公共空间装饰墙面 / B3 public space for wall decorations.

→B4篮球场 / B4 basketball court.

↓施工中空间机具拆除与楼板打开 / Construction of open floor space for equipment dismantling.

↑墙面化身为集团的历史数据库 / Wall incarnation of the Group's historical database.
← 空间设计透视 / Space design perspective.
↓旧有印刷机具空间原貌 / Printing machinery room for the old original.

ASUS DESIGN CENTER

华硕设计中心

12

【坐落地点】中国台湾台北市北投区立德路15号

【面积】1 275 m^2

【设计】龚书章；【设计公司】原相联合建筑师事务所

【主要用材】玻璃、板岩砖、亚克力、杜邦石、木地板

本方案办公室的配置由两条大街联结起正式和非正式的办公空间，在移动的工作环境里共享资讯，激发出更多的创意。空间以黑白为主，由木质地板铺陈。在公共空间，这里营造出像在家里的书房、餐厅一般的亲近。一面面的隔间墙画满涂鸦，几道白色墙面挂满各式各样的员工私人物品。可穿透的大会议室，十扇门片便是与外的屏障，同时推开，不仅可与客厅及吧台区串联，亦表现出空间活泼的韵律感。没有了往日条理清楚的办公室的模样，更多的是像一个微型城市般的热闹和包容。员工在他们的办公室彻底放松刺激了他们的设计灵感，这正是业主想要达到的。

以"微型社区VS家"为设计主轴的华硕设计中心用以人为本的设计概念，结合多元环境设计思维，取代平面单元化的规划方式，并将传统注重管理与效率的制式框架空间，转换成充满互动与合作的开放式工作氛围，获得业界高度肯定。

The program office's configuration consists of two streets to link up the formal and informal office space, in the mobile work environment to share information, stimulate more creativity. Space-based black and white, from wood floors to elaborate. In the public space, where as in the home to create a study, close to restaurants in general. Side surface of partition wall painting over graffiti, with a few white walls covered with a wide variety of staff personal effects. Penetrate the large conference rooms, 10 doors and outside the barrier film is also open, and not only in series with the sitting room and bar area, but also demonstrate a sense of space and lively rhythm. There is no coherent past the appearance of the office, more like a miniature city like fun and inclusion. Employees in their offices completely relaxed stimulate their design inspiration, this is the owners want to achieve.

To "micro-communities, VS home" for the design of the ASUS design center axis for the people-oriented design concept, combined with multiple environmental design thinking, to replace the flat cell-oriented planning approach, shifting the traditional focus on the management and efficiency of the standard framework for space, converted into a full interactive and cooperation, open working environment, access to the industry recognition.

↑休息区 / Rest area.

↑中心办公室 / Central office.
↓平面图 / Plan

↑一面面的隔间墙画满涂鸦 / Side surface of partition wall painting over graffiti.
↓可推拉的门 / You can push and pull the door.

QUANFENG OFFICE BUILDING

泉峰总部办公楼

13

【坐落地点】南京江宁

【设计】黄斌、施亮

【深化设计】上海现代建筑设计（集团）上海现代建筑装饰环境设计研究院有限公司

【摄影】贾方

本方案构思采取大块面组合方式，将建筑、景观和室内综合考虑，同时引入中国园林空间特征要素，使室内与室外取得了良好的对话与交流。

方案强调室内设计在功能布局上的合理性，追求人性化和可持续发展性。在设计风格的思考中，依靠整体氛围的塑造及空间形式与细部的把握，创造现代、实用、高品味的室内空间氛围。在风格与材质上做到主次划分，明确材料之间的交接。材料的运用方面注重大块面的对比效果和细部节点精致推敲的结合，重点强调室内空间的体量感和构造魅力，避免过度地对建筑空间进行刻意的装饰。在室内设计间隔划分上，多采用现代简约、清晰明快的手法，运用柔和的线条与刚毅质感的材料及空间块面的切割。

设计中充分利用自然光和人工光的相互映衬关系，令自然光成为空间辅助照明，使光线效果营造出明亮的、柔和舒适的视觉环境。同时还把握好适当的灯光照明，按照不同空间的需要，凸显出室内结构的轮廓、空间层次和办公家具及装饰物的体积感，改变人们的空间感觉，并打破预先设定的空间分界，突出节奏的变化与空间层次感，使有限的内部视觉空间尽量放大，达到美化空间的效果，创造出新颖的、有吸引力的办公空间。

The program idea to take large pieces of surface composition, the architecture, landscape and interior comprehensive consideration, while the introduction of elements of spatial characteristics of Chinese gardens, indoor and outdoor access to a good dialogue and exchanges.

Interior design program emphasizes the functional layout of the rationality of the pursuit of human and sustainable development. Reflections in the design style, relying on the overall atmosphere of the shape and spatial form and mastery of detail, creating a modern, practical, high-taste and atmosphere of interior space. Done in style and material on the primary and secondary division of a clear interface between materials. Focus on the utilization of materials, large surface contrast effects and detailed scrutiny exquisite combination of nodes, with emphasis on the amount of interior space a sense of the body and structure charm, to avoid excessive space on the building deliberately decorative. Intervals in the interior design division of multi-use of modern and simple, clear and crisp way, using soft lines and textures of materials and space fortitude block cutting surface.

Designed to take full advantage of natural light and artificial light silhouetted against the mutual relations, natural light into space auxiliary lighting, so that lighting effects to create a bright, soft and comfortable visual environment. Is also a good grasp of appropriate lighting, according to the needs of different space, highlights the contours of interior architecture, space-level and office furniture and decorative items in the volume of a sense of space change the way people feel, and the space to break the pre-set boundaries, highlighting a change of pace and space-level sense, so that the limited visual space as far as possible within the amplification.

↑室内局部 / Interior space.

↑总部办公楼 / Headquarters office building.
← 楼梯 / Stairs
↓简洁、明快、结构清晰的室内 / Concise, crisp, clear indoor structure.

098

↑主入口门厅设计 / Design of the main entrance hall.
→行政餐厅 / Executive Restaurant.
↓设计中充分利用自然光和人工光的关系 / Make full use of the relationship between natural light and artificial light.

100

↑室内大堂局部 / Interior lobby local.
↓会议室走廊 / Meeting room corridors.

↑通道 / Channel
↓旋转楼梯 / Rotate the stairs.

UTOP SCIENCE AND TECHNOLOGY OFFICE

UTOP优伯科技办公楼

14

【坐落地点】厦门同安工业集中区思明园7号

【设计】徐福民

【主要材料】钢结构、轻钢龙骨硅酸钙板、钢化玻璃、地面水洗石、墙面乳胶漆

【摄影】吴永长

整体来看，全部空间就像是块、面和玻璃构架起来的一个透明盒子。设计师以数片较薄的墙体立于盒子之内，又以较厚的墙体形成穿插其中的块，二者互为转折传承，不仅控制了室外光线的进入程度，同时也形成了室内与室外之间的灰空间，透过玻璃，望着外面的绿树和天空，经过过滤的柔和天光流淌在盒子的墙壁上，室内室外已然模糊了界线，达到了设计师的本意。

信息时代的解构主义是这个设计的特点，空间和几何结构的大胆运用，反映出现代信息社会积极、动感的特质，给人以高科技的联想。从空间的色彩来说，我们可以体会到设计师深厚的中国传统文化素养。深味道教文化的设计师在设计中采用近乎无色的设计，简单清爽的灯光散发出一阵阵安静的气息，让人安心于此，浸意思维。西方现代艺术和中国传统道家思想在这得到了融合。

Overall, all the space like a block, face and glass architecture up a transparent box. Designer in order to thin the wall stood a few pieces inside the box so, on the formation of a thick wall blocks interspersed among them, the two pass each other turning point, not only control the entry of outdoor light levels, but also formed a house between the gray and outdoor space, through the glass, looking out of the trees and sky, the soft daylight flowing through the filter walls of the box, indoor and outdoor already blurred boundaries, to achieve the designer intended.

Deconstruction is the information age, the characteristics of this design, space and the bold use of geometry, reflecting the modern information society, active, dynamic characteristics, giving high-tech associations. Color from space, we can appreciate the profound traditional Chinese cultural qualities designer. Deep taste of the designer in the design of teaching culture used in the design of near-colorless, simple and refreshing waves of light emitting a quiet atmosphere, people feel at ease here, the Baptist Italian thinking. Of modern Western art and traditional Chinese Taoism has been integrated in this.

↑空间就像是块、面和玻璃构架起来的一个透明盒子 / Space like a block, face and glass architecture up a transparent box.

↑办公空间 / Office space.
↓创的楼梯造型 / Creating a staircase shape.

↑灯光效果让空间变得温暖 / Lighting effects make room to become warmer.
↓一层平面图 / Floor plan.

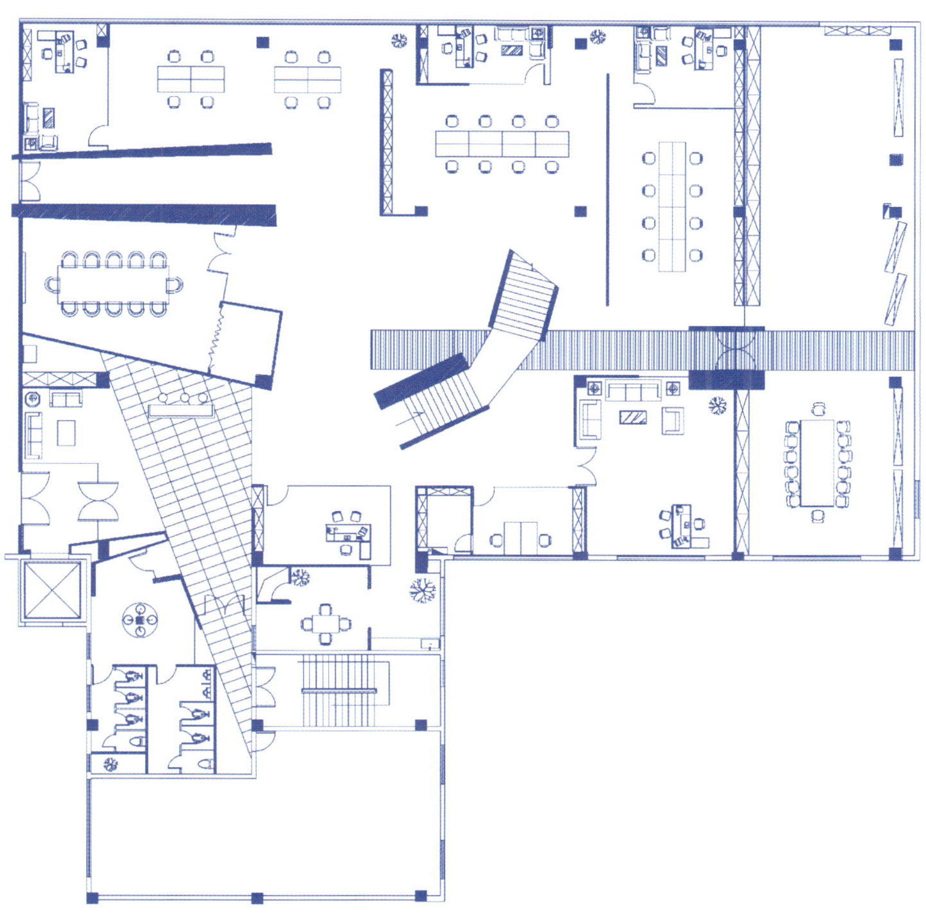

LINGUAPHONE LANGUAGE TRAINING

灵格风语言培训机构

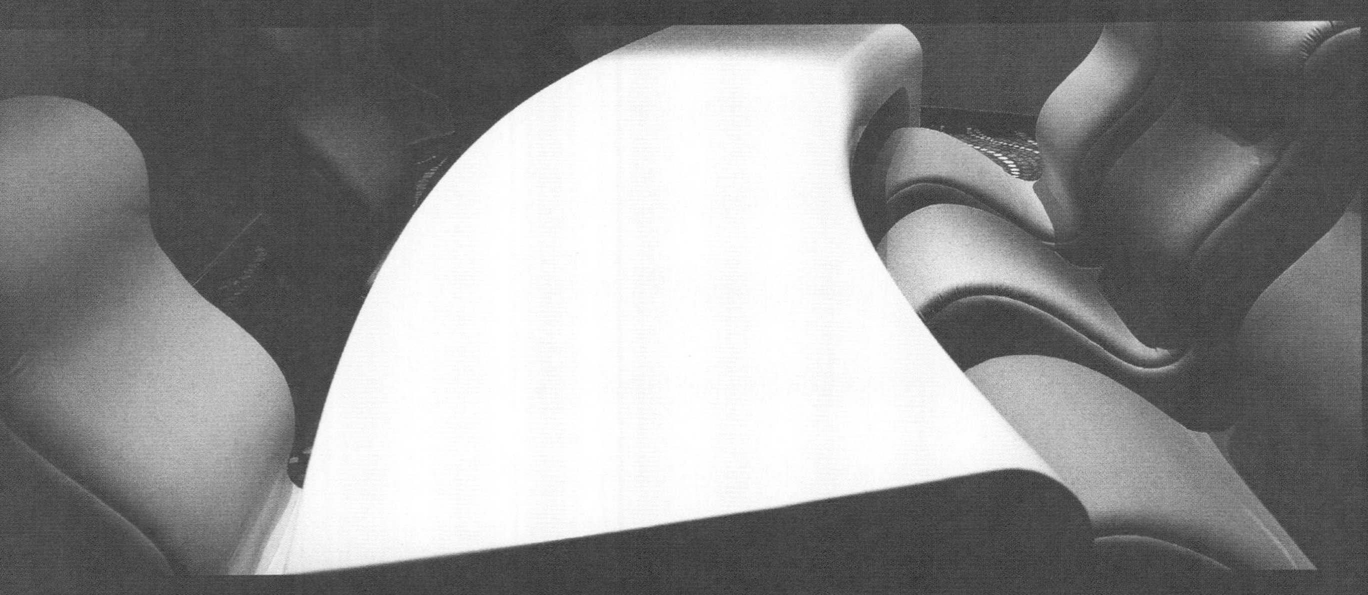

15

【坐落地点】广州越秀区农林下路；【面积】约2 000 m^2

【设计】黄志达；【参与设计】李家贤

【主要用材】PU高光漆、灰木纹大理石、阿加力胶版、马赛克

【摄影】陈思

灵格风(LINGUAPHONE)，一个风靡全球的著名语言培训机构。广州培训中心的空间设计秉承了时尚、简约、优雅、自由自在的整体风格，目的是打造一个国际活动式的学习中心，主要功能是学习+娱乐，所以我们并没有按照正规学习场所那样布置得规整而传统，相反采用了一些比较轻松的表现手法。

入门处的接待台及接待区域，稳重的实木隐隐展露出自然的纹理，"LINGUAPHONE"在浅浅的灯槽中披着柔和的光芒，很素雅很亲近，带给每一位来这里的客户一份值得信赖的平静的感觉。

休闲区的接待台完全有别于前台接待的素雅，斑斓的LED泡泡灯装扮背景墙，不规则的色彩搭配，随意而轻松自在，一个经意摆设在接待台一角的小公仔，是灵格风的形象大使，大大的黑溜溜的眼睛，神采奕奕，吸引了活泼的小朋友们。

转至洽谈区，重返沉稳的淡雅色调，舒适静寂，让每一次交谈都成为情感的交流，共同为孩子们规划最有效的学习方式，创造一片更自在的学习氛围，真实而值得信任。

LINGUAPHONE is a rage, renowned language training institution. Guangzhou training center space design uphold fashion, simple, elegant, free to the overall style, the purpose is to create an international activity-based learning center, the main function is to study and play, so we have not done in accordance with informal learning spaces decorated neat The traditional, on the contrary used the performance of some of the more relaxed approach.

Getting Started at the reception tables and reception area, prudent faint exhibition revealing the natural wood texture, "LINGUAPHONE" dressed in the shallow soft light bath light, very simple and elegant is very close to come here to bring every customer a were trusted calm feeling.

Leisure areas, completely different from the Reception Desk Reception of the simple and elegant, gorgeous dress backdrop of the LED bubble lights, irregular color match, casual and comfortable, a Jingyi decorations at the reception corner of a small doll, yes Linguaphone Wind ambassador yo big black eyes, attracting lively kids.

Go to negotiate areas, return to the steady tone of elegant, comfortable and quiet, so that every conversation becomes emotional sharing, joint planning for children to learn the most effective way to create a more comfortable learning environment, a true and worthy of trust.

↑培训中心走廊，斑斓的色彩，流动的曲线 / Training Center corridor, gorgeous color, flow curve.

↑休闲区接待台 / Recreation Area Reception Desk.
↓舒适的沙发 / A comfortable sofa.

↑阅读区的大红半圆沙发 / The red semi-circle sofa reading area.

↓平面图/ Plan

↑模仿太空的学习室 / Imitation space learning centers.

↓放松区域 / Relaxation area.

↑色彩跳跃的学习室 / Color jump study rooms.
↓洽谈区色调淡雅，舒适静寂 / Discuss areas elegant colors, comfortable silence.

ONE PLUS PARTNERSHIP

壹正企划办公室

16

【坐落地点】中国香港北角英皇道
【工程名称】壹正企划办公室
【设计】罗灵杰、龙慧祺
【摄影】罗灵杰、龙慧祺

作为一间设计公司的办公空间，其风格完全体现了设计师们的喜好，黑色的大门开启了设计师深层的思维空间，迎接或者说引领客人进入设计师创想世界中。玄关氛围昏暗，走入其中即被一种奇特的情绪所包围，黑色特式墙上的石头漆和先锋派字体的公司招牌，简明扼要地呈现了该公司非同一般的设计笔触。再往里面走则是另外一番“柳暗花明”，闪绿色布帘和绿色古典花纹坐椅，与附近的环境形成光暗的对比，天花上错落有致的灯光打在空间中的各种物件上，营造出仿佛戏剧舞台般的场景感。

不同于一般公司里庄重严肃的会议室，设计公司的会议室带给人眼前一亮的感觉。光亮的爵士白云石会议台和造型前卫的坐椅组成有趣的搭配，显示着当代年轻人勃勃的生机；轻巧的窗帘挡住了室外的天光和英皇道繁华的街景，让设计师在进会议时免于外界不必要烦扰。

处于后台的工作区简约而有序，是值得一提的特式设计，该处的分区以开放式间隔完善地划分出来，设计师最喜爱的黑色依然是当中的主调，除此之外再用适当的颜色缓和了黑色的吸光特性，调和了空间中的色彩和光线比例。最为夺目的设计是工作区内订制的十字形灯管组合，在暗沉的天花背景下，它们以独特的造型和明亮的光线变得更加耀眼。同时，十字造型与公司的标志有着异曲同工之处，时刻强调着公司的团结协作的工作理念。

As a design company's office space, its style, completely embodies the designers preferences, black designer opened the door to deeper space of thinking to meet or to lead a guest into a designer would like to create a world. Vestibule atmosphere of dark, into a strange mood which Jibei surrounded by a black Gothic stone wall paint and avant-garde font company signs succinctly presents the design of the company's extraordinary general strokes. And then puts inside the big picture to go is another, flashing green curtains and green classic pattern chair, with its surrounding environment the formation of brightness contrast, smallpox hit the lights on a patchwork of various objects on the space, creating a the stage-like scene as if a sense of drama.

Unlike most serious and solemn company meeting rooms, designed to bring the company's conference room were shines feeling. Light jazz avant-garde dolomite conference table and chairs form an interesting mix of the composition, showing the vitality of contemporary young people thriving; lightweight curtains blocking the outdoor daylight and King's bustling streets, so that designers into the meeting unnecessary annoyance from the outside world.

Unlike most serious and solemn company meeting rooms, designed to bring the company's conference room were shines feeling. Light jazz avant-garde dolomite conference table and chairs form an interesting mix of the composition, showing the vitality of contemporary young people thriving; lightweight curtains blocking the outdoor daylight and King's bustling streets, so that designers into the meeting unnecessary annoyance from the outside world. Relaxed and comfortable environment conducive to promoting communication and cooperation, creative sparks burst out naturally.

↑玄关的设计极富舞台的戏剧性 / Vestibule design of highly dramatic stage.

↑从玄关通往办公区的过道 / From the vestibule leading to the office of the aisle.
← 绿色的布帘和古典花纹坐椅带来一抹亮色 / The green curtains and classical patterns to bring touch of light-colored chairs.
↓工作区，很平常的空间组合 / Work area, it is the usual space combination.

↑设计总监办公室 / Design Director's Office.
← 简约而轻松的会议室 / Simple and easy conference room.
↓平面图 / Plan

BURATTI + BATTISTON

B+B建筑事务所

17

【坐落地点】 Busto Garolfo

【面积】一层200 m²，**第二层**100 m²，**仓库与模型间**100 m²

【设计】 Buratti + Battiston Architects；**【参与设计】** Gabriele Buratti, Oscar Buratti, Ivano Battiston

【摄影】 Andrea Martiradonna

木工厂建造于上个世纪50年代，被一片树木和田地包围着。事务所所在的部分原来是木工厂的商店部分。这是一座矩形的建筑，有一个拱顶。结构和周围的环境都很简单。建筑一度外加了个斜顶，这次重新设计时把它去掉了，又恢复了原来的面貌。

一层入口处设置了大堂和接待处，一个小型的会议室和卫生间。紧靠着楼梯的部分是一个共享的两层高的工作室，空间比较高大适合布置成工作区域。二楼有一个咖啡间，一个大型的会议室也兼做图书收藏室，当然也有一部分工作区域。建筑中心的拱顶下方一部分是两层结构，其余是共享的部分。二层既能起到调整空间的作用，安排一些功能区域，更重要的是把灯光藏在里面，保证整个建筑的照明。

连接两层的楼梯间是整个设计的亮点。这是个由玻璃组成的透明"盒子"。所有的踏步采用了表面没有被精打磨过的，粗加工的混凝土钢板，从玻璃外面能一眼看到它们。这些踏步是由激光把整块混凝土钢板截断而成，表面笔直硬朗的线条和横截面的曲线形成了鲜明对比。材料的选择则使得楼梯比以前更加突出。这样的设计让楼梯看上去简洁大方，轻盈闪亮。上下两层的地面材料不大一样：一层采用了颜色较浅、光泽度较好的复合地板，二层则采用了较厚重的橡木地板。为了保持工作环境的严肃性，办公室的墙、天花和家具基本上都很简单。

Wood factory was built in the last century 50's, was one surrounded by trees and fields. Firm is located was originally part of the store part of the wood shop. It is a rectangular building, there is a vault. Structure and the surroundings are simple. Building once a month plus a slant, this re-design it removed, but also to restore the original appearance.

Level set at the entrance lobby and reception area, a small conference room and toilet. Close to the staircase is part of a shared two-storey studio, space for tall and fit into the work area layout. On the second floor has a coffee room, a large conference room also cater to book collections, and of course part of the work area. Architecture Center of the vault beneath the part of a two-tier structure, and the rest is shared part. Two-story space both play a role in adjusting to arrange a number of functional areas, but more importantly is to light hidden inside to ensure that the whole building lighting.

Staircase connecting two floors of the highlights of the entire design. It was composed by a glass-transparent "box." Use of all of the stepping surface has not been refined polished, and rough concrete steel, from the outside can be a glass to see them. These stepping from the laser to cut steel plate made of concrete block surface tough straight lines and cross-section curve is in sharp contrast. The choice of materials was to make the stairs more prominent than ever before. This design allows the staircase looks simple and generous, light flashing. Ground up and down two quite different materials: one using the color lighter, better gloss composite floor, second floor while using a more heavy oak flooring. In order to maintain the solemnity of the work environment, the office walls, ceilings and furniture are basically very simple.

↑ 纵横的书架 / Aspect of the bookshelves.

← ↑一层办公区 / Floor office area.
↓一层平面图 / Floor plan.

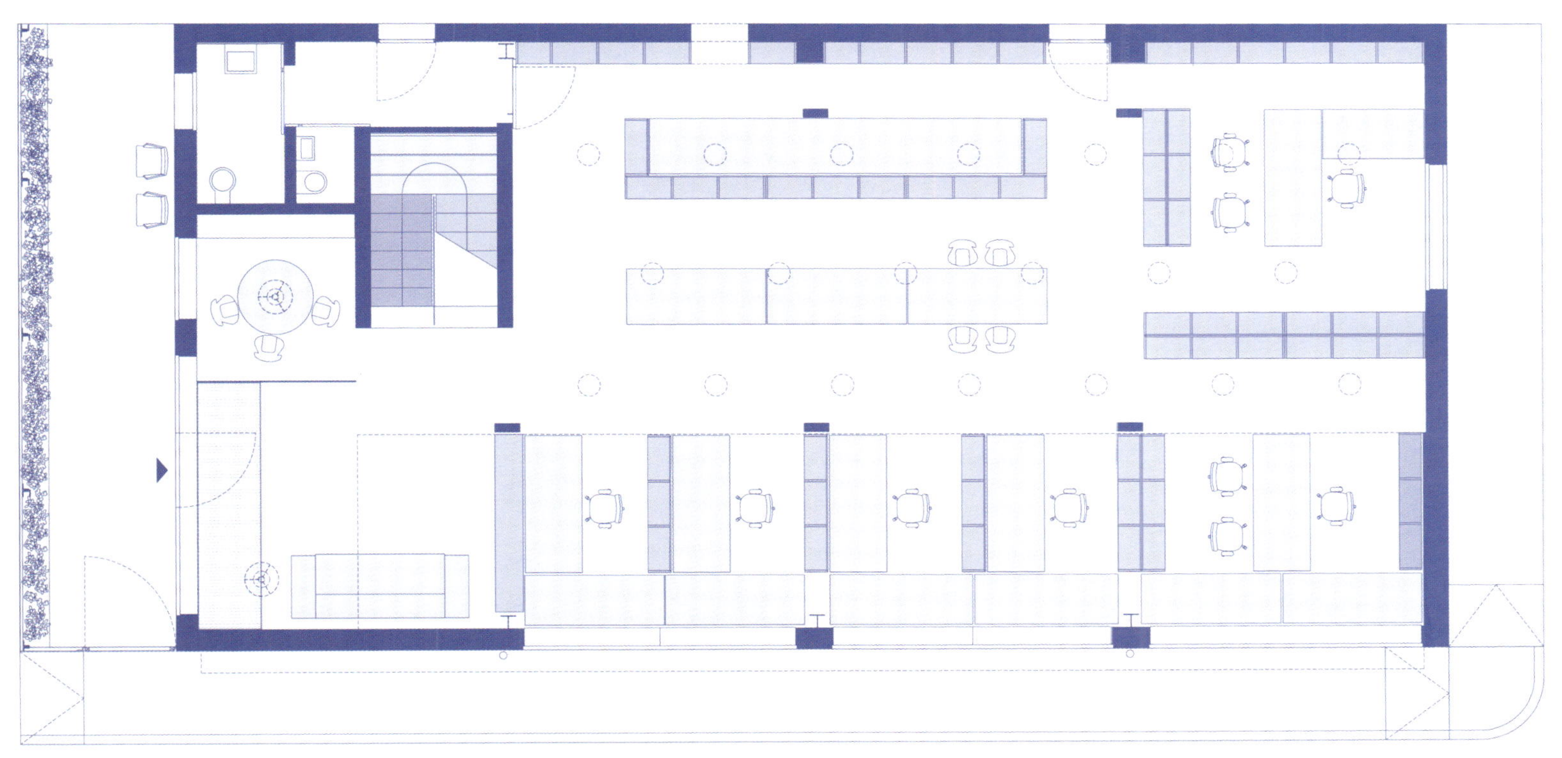

←↑ 从窗外看办公室 / From the window to see the Office.
↓ 二层平面图 / 2nd floor plan.

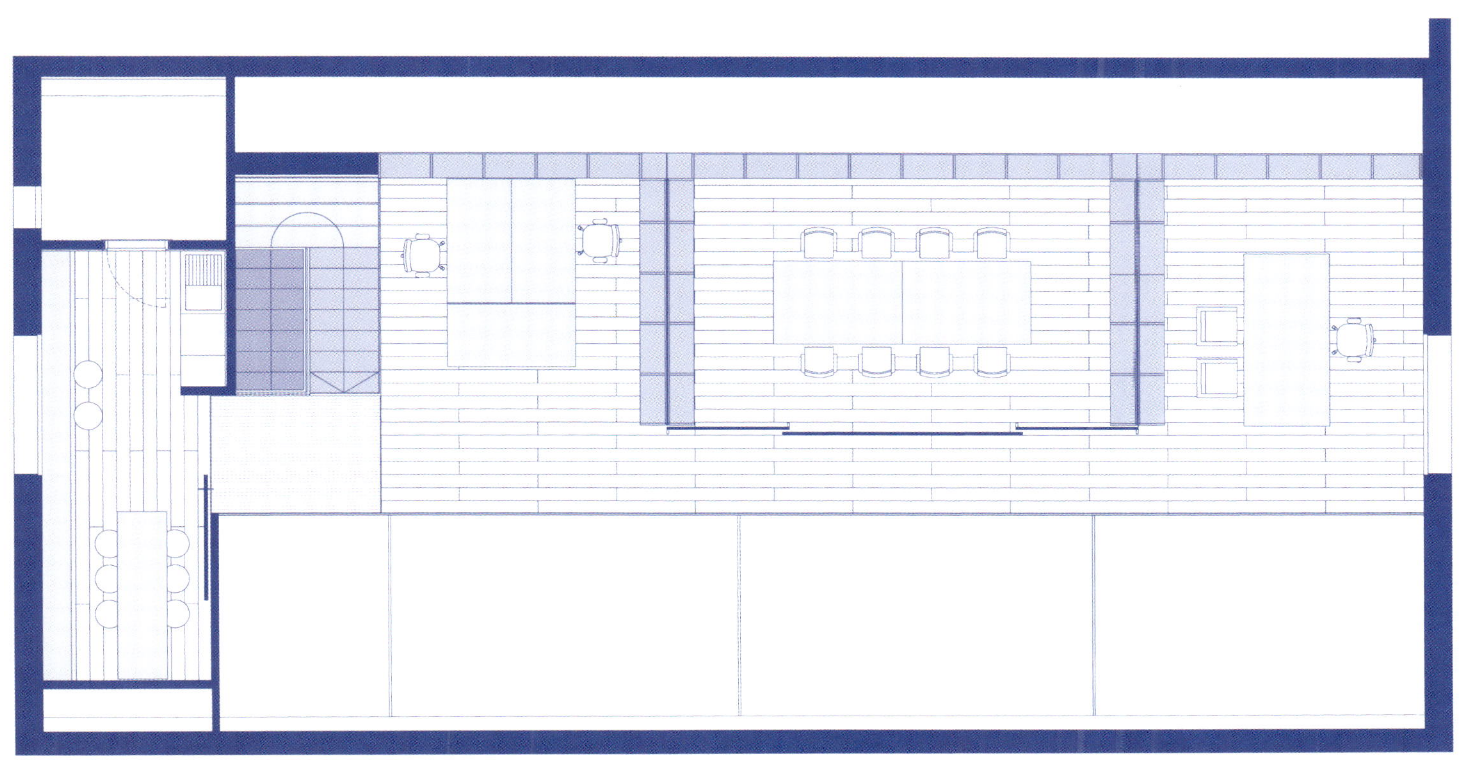

↑↑楼梯对面的休息区，沙发颜色很鲜艳 / Staircase opposite the rest area, sofa, color is very bright.
← 二层的大会议室 / The second floor large conference room.
↓三个方向的立面图 / The elevation in three directions.

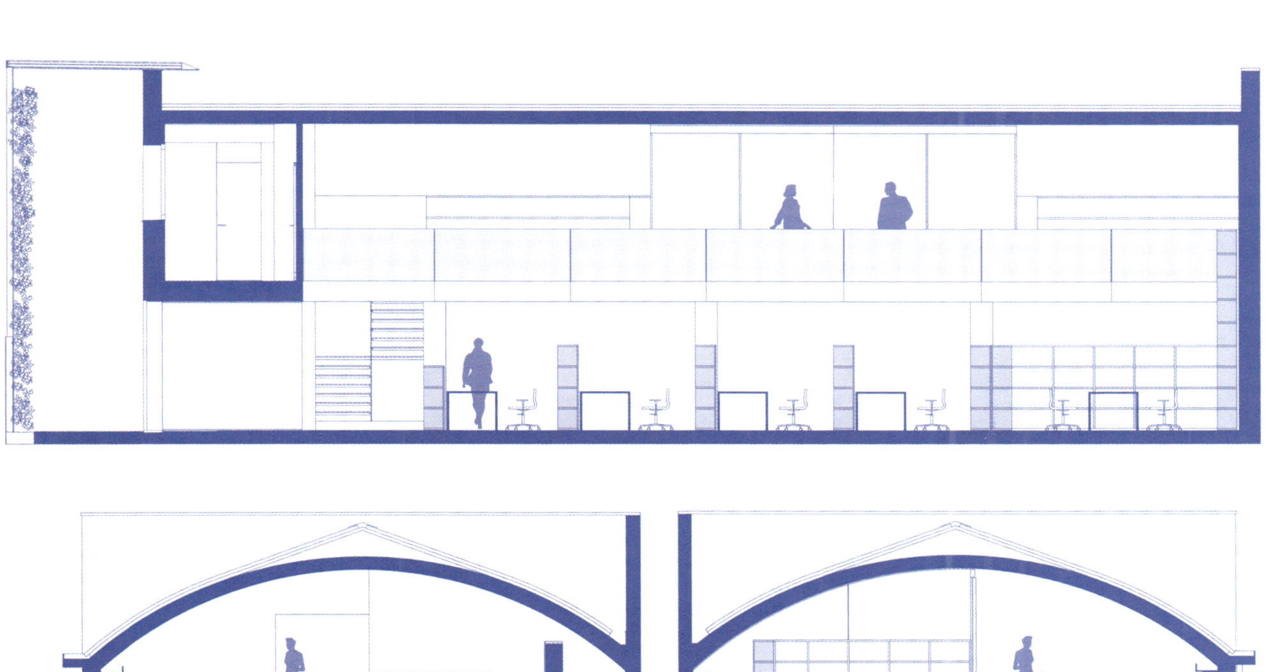

ROBARTS INTERIOR AND ARCHITECTURE

一工室内设计建筑公司

18

【坐落地点】北京CBD商务区写字楼
【工程名称】一工室内设计建筑公司
【面积】829 m^2
【设计】北京一工室内设计建筑公司

作为一家专业从事办公空间设计的公司，他们自己的办公室设计当然是最能够体现其专业水准和设计理念的作品。走进公司，你会发觉这里的氛围高贵典雅，散发着与众不同的非凡气质。公共活动区域挑高6 m，既体现了这一区域的核心地位又让身处其中的人们感觉到别样的舒适惬意。倾泻而下的影壁把公司的LOGO恰当的展现在了访客面前。古典和现代气息并存的休息大厅，为员工们提供了可以随意交流、读书和休憩的场所。通向二层的两处楼梯都伴有简约而别致的水景，潺潺流动的清水为整个办公空间带来了无限生机，这流水声又似舒缓的乐曲让员工在工作之时心境更加的放松。透过巨大的落地玻璃窗更可以看到宽敞的户外屋顶平台上的绿色草坪，在北京这样高楼林立的商务写字楼区，这当然是罕见的景致。

这间办公室员工的办公区与客户接待区之间的界限是模糊的。注重工作区域的人性化设计是设计师的宗旨。在这里你甚至还能找到有两个独一无二的空间——冥想室和按摩室。冥想室是员工们天马行空放松心情的地方。在按摩室里，大家每周都能享受到来自专业按摩师的特殊呵护。

As a professional office space design company, their own office design is best able to reflect their professional standards and design work. Into the company, you will find the atmosphere here, noble and elegant, distinctive exudes extraordinary temperament. Public activity area with a ceiling 6 m, not only embodies the central position of letting the region in which the people living in different kind of a safe and comfortable feel. Pour down the screen wall to the right shows the company's LOGO in front of the visitors. The coexistence of classical and modern lobby, employees are free to provide a communication, reading and sitting place. The two staircases leading to the second floor are accompanied by simple and unique water features, unimpeded flow of water has brought the entire office space full of life, it also seemed to ease the flow of music sound of the staff time at work more relaxed state of mind . Floor through the large windows we can see the spacious outdoor roof terraces on the green lawn, the tall buildings in Beijing, such a business office area, which is of course a rare scene.

This office employee office area and reception area, the boundaries between clients is ambiguous. Focus on the humanization of work area design is the designer's purpose. Here you can even find two unique space - meditation room and massage room. Meditation room is the employees were unrestrained relax place. In the massage room, everyone can enjoy a week of special care from professional masseurs.

↑金属与玻璃组合的栏杆干净而通透 / Combination of metal and glass railings clean and transparent.

↑↑简洁大方的设计体现出典雅的美感 / Simple elegant design reflects the elegance and beauty.
← 古典和现代气息并存的休息大厅 / The coexistence of classical and modern lobby.
↓↓简约时尚的楼梯 / Simple stylish staircase.

↑↑办公区域和交流区的界限是模糊的 / The exchange of regional and district office boundaries are blurred.
← 二层的大会议室 / The second floor large conference room.
↓↓独特的细部设计 / Unique detail design..

PUBLICMOTOR BRAND COMMUNICATION

某办公空间设计

19

【坐落地点】德国斯图加特
【面积】700 m^2
【设计】Bottega and Ehrhardt Architekten〔德〕
【摄影】David Franck

本案以大面积的白色为主要色彩基调，搭配以灰色、黑色显示了办公空间专业、严肃的一面，而明亮黄色的色彩点缀则是画龙点睛之笔，霎时间让气氛活跃起来。办公空间的照明设计也非常的讲究。由于办公时间几乎都是白天，设计师特别将人工照明与天然采光结合起来，并且采用柔和的漫反射光源。在进门处，设计师还在天花板上设计了一些光影的变换，来增添室内的趣味性。

从进门处地面开始的金色圆圈在等待区内变大，巨大的圆圈内摆设了两个充满现代感的沙发椅，它们的设计新颖现代，让人们感受到这里不一样的氛围。在此坐着等待的时候，便可以注意到前面那些三角体的秘密，它们都被植入了小的LCD液晶显示器，上面不停的展示着PUBLICMOTOR过往的工作和成绩。这些立方体一物多用，分割空间、有效隔音，展示作品的同时，又不影响整体照明的效果。公共办公区域的设计开放流畅，中间用低矮的文件柜相隔。在开放办公室的尽头，是职员休息室。它们同时也把办公区与公共交流区隔开，长长的木质桌子保留了木材原有的纹路，清新自然。

简洁大方、注重功能是本案的一大特色，设计师没有将设计重点体现在各种变化的手法上，而是更注重办公空间的舒适度，最大限度地满足了人性化和功能化的需求，也是现代设计中最为重要的模式之一。

The case of large areas of white as the main color tone, with gray, black, with office space for professional and serious side, while the bright yellow color is very eye-catching embellishment, and all of a sudden make the atmosphere become active again. Office space is also very particular about lighting design. Since almost all hours during the day, designers specifically artificial lighting and natural light combine, and the use of soft diffuse light. In the entrance, the designer also designed a number of light and shadow on the ceiling of the transformation, to add the indoor fun.

From the entrance on the ground waiting for the beginning of the golden circle area bigger, huge circle filled with modern furnishings, two sofa chairs, their modern design modern, so that people feel here, not the same atmosphere. Sitting in this waiting time, it may be noted that these triangular body in front of the secret, they have been implanted into the small LCD liquid crystal display, the above non-stop display of the PUBLICMOTOR past work and achievements. The cube a complex multi-purpose, divide space, effective sound insulation, showing works at the same time, without affecting the overall lighting effect. The design of public open office areas smooth, the middle interval with the low filing cabinet. In an open office at the end, is the staff lounge. They also share the office and public areas are separated by a long wooden table and retain the original texture of wood, fresh and natural.

Simple and generous, pay attention to function is a major feature of the case, designers did not design focus is reflected in various changes in the way, but rather focus more on office space, comfort, to best satisfy the user-friendly and functional needs, is also a modern design, one of the most important patterns.

↑办公室入口 / Office of the entrance.

↑接待区的金色的圆圈内摆设了两个充满现代感的沙发椅 / The reception area of the golden circle filled with modern furnishings, two sofa chairs.

↓公共办公区域的设计开放流畅 / The design of public open office areas smooth.

↑大面积白色的应用，加上黄色的点缀让空间立即鲜活起来 / The application of large area white, coupled with the yellow dotted room immediately live up to.

↓平面图 / Plan

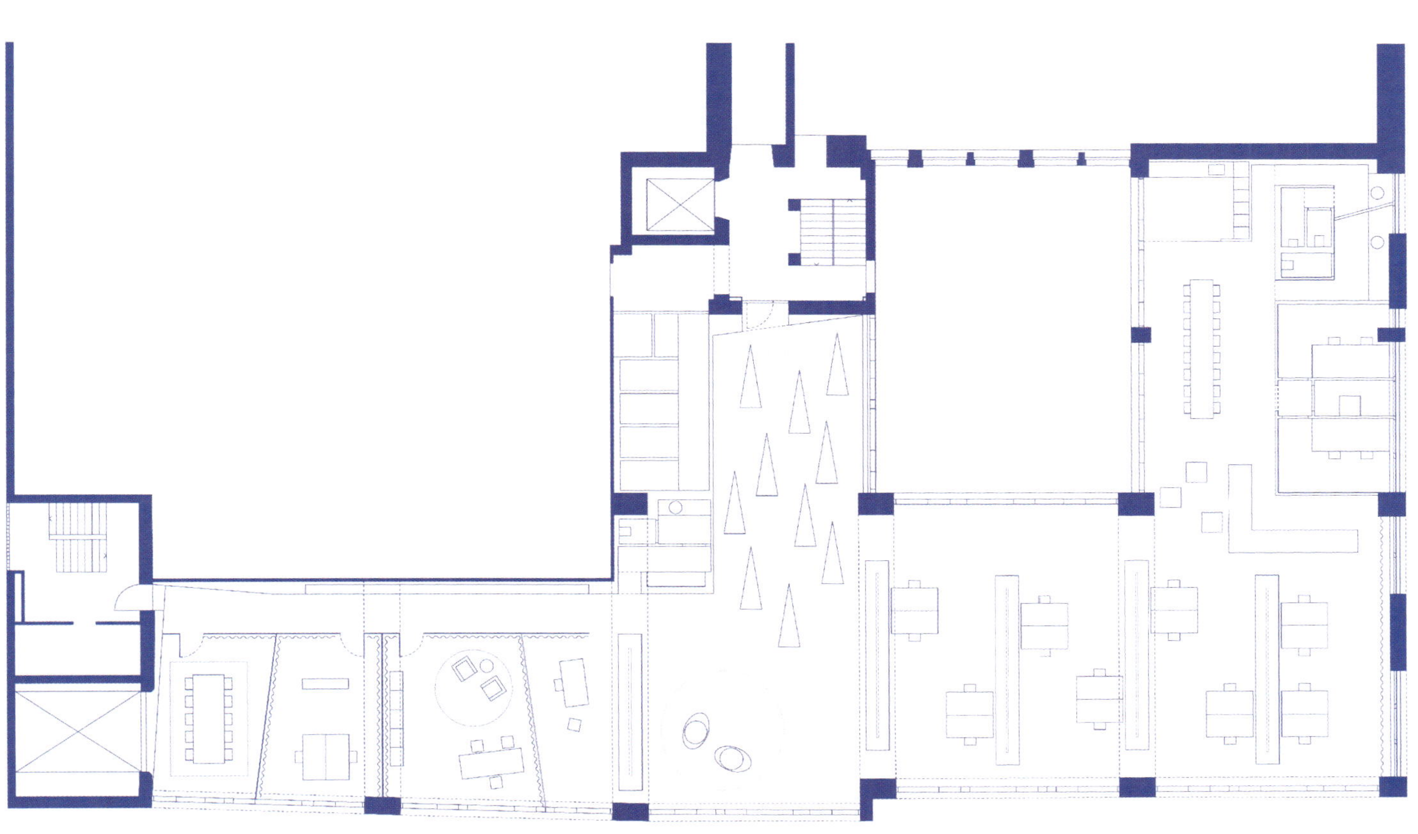

↑黄色柜子的书架，让休息室成为一个临时的小型图书室 / Yellow cabinet's shelves, so that a small lounge into a temporary library.

↓这里连着一个小型的厨房，供大家中午午餐使用 / Here attached a small kitchen, lunch at noon for everyone to use.

↑设计师使用青铜色的帷幔来保证工作场的私密 / Designers using the bronze curtain to ensure the privacy of the workplace.

↓办公空间力求最大限度的满足人性化和功能化的需求 / Office space, an effort to maximize the satisfaction of human and functional needs.

CROCS OFFICE IN SHANGHAI

Crocs上海办公室

20

【坐落地点】上海西康路300号；【面积】1 800 m²
【设计】颜呈勋
【主要材料】超白钢化玻璃、亚克力、铁板冲孔钢板
【摄影】申强

在这个案例中，设计师摆脱了惯常的局部封闭办公室与大面积开敞办公相结合的二元办公空间设计模式，而是将所有的职能部门与功能区块分散重组，有机结合，从而形成了多元高效的办公模式。

基于CROCS业主对整个公司的流程的考虑，MRT的设计师在此要求上，对于办公附属功能空间的进行了巧妙的布置。例如，合理利用写字楼核心筒位于整个楼层平面中央这一特点，将日常使用频率较高的共享办公区域都集中布置在核心筒附近，从而获得更加高效的空间利用率。

色彩的搭配与运用亦是经过设计师深思熟虑的结果。在这个项目中，设计师运用了大量明快的色彩，而这些彩色涂料的品质与环保特性就是设计师必须面对的问题。MRT的设计师选用了一种产于荷兰的环保装饰涂料，无需通过时间或通风来驱散气味，将对使用者的危害降至最低，体现出设计师对于空间使用者的关怀与考量。

设计师亦通过办公家具的选用与设置，体现出其自始至终对于设计的重视。在该项目中，设计师放弃了传统的三面围合的办公隔间，而在整个办公空间中，从高层领导到基层员工，统一采用直板式工作站。用半透明玻璃作为工作站的隔断屏风，屏风的高度确定在某一高度，使得相对而坐的人独立工作的同时，也可以直面交流。

In this case, the designers from the usual partial closure of offices and large open office area of a combination of office space design pattern, but all the functions and function blocks scattered reorganization, combination, creating a diverse and efficient office model.

In order to meet the needs of the owners, designers for office space for the conduct of the subsidiary functions of the clever layout. For example, the rational use of office floor flat core tube is located in the central government this feature will be a higher daily frequency of use of shared office space is concentrated near the core tube arranged in order to obtain more efficient space utilization.

Color matching and application designers are also subject to the results of careful consideration. In this project, the designers used a lot of bright colors, and these color paint quality and environmental features that designers must face. MRT designers opted for a production of environmentally friendly decorative paints in the Netherlands, without the need for time or ventilation to disperse the smell, it will harm the user to minimize space for users to reflect the designer for the care and thought.

Designers also endorsed the selection of office furniture and settings, reflecting their importance for the design from start to finish. In this project, designers abandoned the traditional enclosed on three sides of office cubicles, while the entire office space, from the senior leadership to junior staff, uniform application of straight plate workstations. With translucent glass partition as a workstation screen, screen height to determine the appropriate height, making relatively independent work of people sitting at the same time, you can also face communication.

↑办公室入口 / Office of the entrance.

↑办公室空间的流畅性 / Office space fluency.
←色彩分布平面图 / Color distribution plan.
↓平面图 / Plan

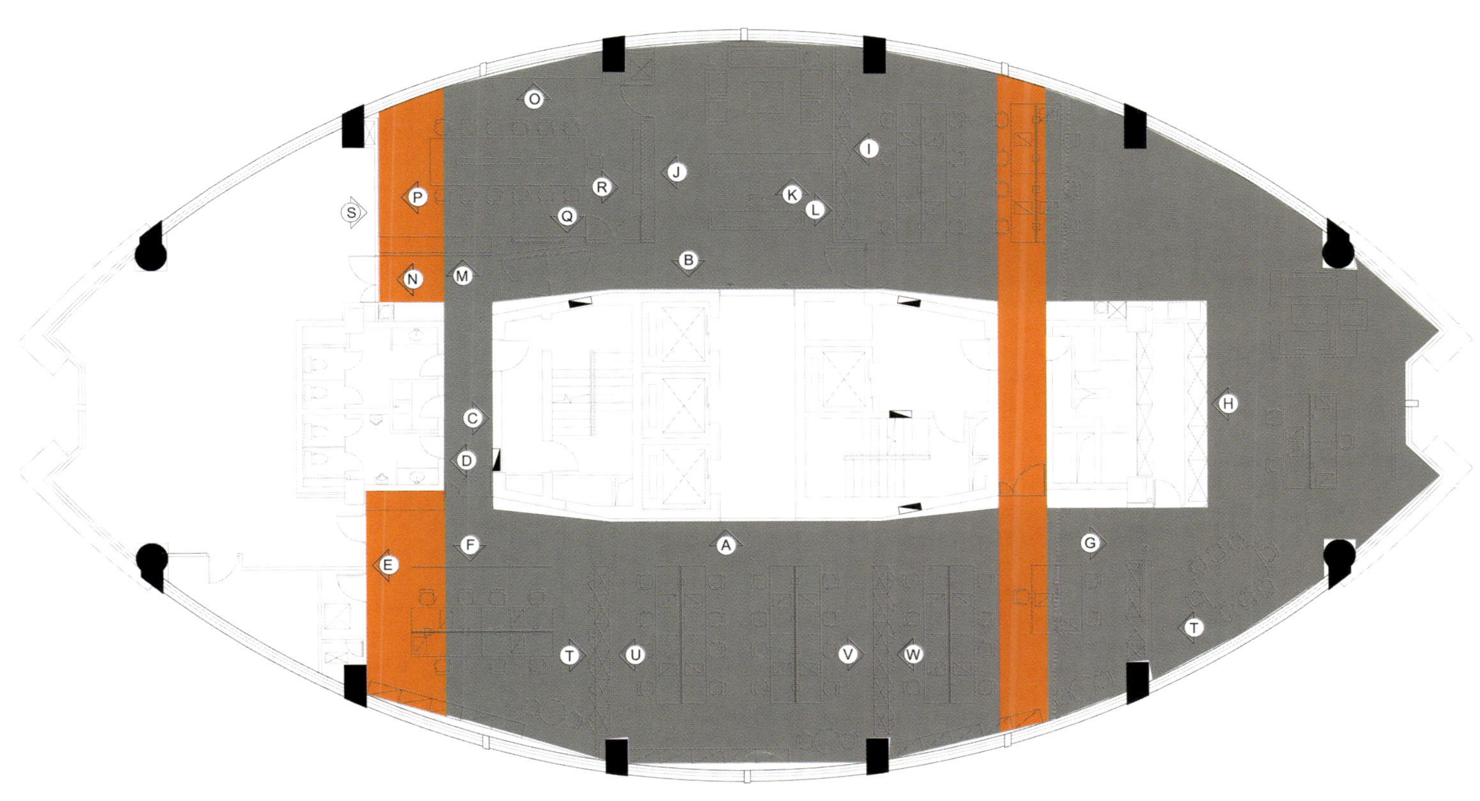

↑工作台屏风的高度适中 / Workbench screen height moderate; ↑灰黑的空间 / Dark gray space.
← 品牌内涵在色彩中体现 / Brand Connotation reflected in the color.
↓休息区 / Rest area; ↓会议室 / Conference room.

AN OFFICE IN BEIJING

建邦矿产资源有限公司

21

【工程名称】北京TOPIRON建邦矿产资源有限公司
【坐落地点】北京银泰中心；【面积】560 m^2
【设计】陈宪淳、石玉冰
【摄影】孙翔宇

设计师通过和业主的沟通后，达成设计的共识是：稳重的、新颖的、时尚的、简洁的。由于这个空间的功能要求非常简单，平面图因此也就没有特别的难度，一两个回合也就定下来。而异常简单的平面也给效果的实施带来新的考验，很多的时候，越简单的格局就越难出好的效果。

前厅的面积占了很大的比例，所以前厅效果的考虑也就在整个方案中占了很大的比例。必须有非常鲜明的印象让访客和内部员工直接感受到公司的全新定位。我们决定用暖灰色来作为整个前厅的主色调，而既然抛弃了木材，那么就再彻底一些——抛弃石材，这个决定没有那么难下，因为有太多的东西可以代替石材，比如说砖。

天花的处理也部分用了镜面的材质，这里是深灰色的镜面不锈钢，视觉上增强了层高，而且暗的颜色又有效地抑制了整个空间"飘"的感觉。天花的另一半做成一个极度简洁的长方形灯槽造型，它与展示柜的背光槽一起组成前厅空间最主要的照明光源，均匀而柔和。

前台就继续延续极简的做法，整体纯白色的人造石，左右稍稍向里张开，配合镜面不锈钢的内嵌式踢脚，加上暗藏灯光，整洁而轻盈。而LOGO字则抛弃所有的颜色，采用镜面不锈钢的材质，与整体氛围和谐共处，完美融合。

Designers and owners through the communication and reached consensus on the design are: stable, innovative, stylish, and compact. As the functional requirements of this space is very simple plan so there is no particular difficulty, one or two rounds also be laid down. The unusual effect of a simple implementation of plane also brought new challenges, a lot of time, the simpler the pattern the more difficult it out good results.

Lobby area accounted for a large proportion, so also consider the effect of lobby throughout the program accounted for a large proportion. Must have very clear impression to visitors and internal staff directly felt the company's new location. We decided to use warm gray as the entire lobby of the main colors, but since abandoned the timber, then the total number of re - abandoned stone, this decision is not so difficult, because there are too many things that can replace the stone, such as brick .

The treatment of smallpox in part by the mirror material, here is the dark gray mirror stainless steel, visually enhanced layer high, and the dark color has also effectively inhibited the entire space, "Gone with the Wind" feeling. The other half of the ceiling made of an extremely simple rectangular light tank model, it is the backlight and display cabinet slots lobby space together make up the main light source, uniform and gentle.

Prospects continue to follow the minimalist approach on the whole pure white artificial stone, or so slightly to the inside open, stainless steel mirror with built-in kick, coupled with hidden lighting, clean and light. LOGO is the word to abandon all of the colors, use of mirror stainless steel material, with the overall atmosphere of harmony and perfect fusion.

↑入口等候区 / The entrance waiting area.

↑LOGO墙 / LOGO wall.
↓暖灰色为前厅的主色调 / Warm gray for the hall's main colors.

↓平面图 / Plan

↑开放办公区域 / Open office space.
← 茶水间的吧台配合休闲沙发 / Between tea bar with casual sofa.
↓会议室 / Conference room.

IADC DESIGNERS

涞澳设计公司办公室

22

【坐落地点】上海市宝庆路20号4号楼B楼
【工程名称】涞澳设计公司办公室
【设计】张成喆
【摄影】路明鑫

设计师只将一片炫目纯粹的白色世界横陈在你的面前。整个办公空间是由白色水泥墙面和长方形立柱分割而成的以直线为主的几何体，没有一处弧线的造型。设计者用最丰富也是最本源的建筑要素——墙体——创造这整个空间。这些墙体和柱体的表面也没有一片瓷砖和墙纸，有的墙面甚至就随便涂抹了些白色的石灰水泥，砖石的缝隙还清晰可见，“计白当黑”，这些立柱墙体让这虚空的白色，在有与无，虚与实，之间灵动地转换。优雅的玻璃立面穿插其间，映着白色的内外墙，阳光自上而下洒落，在纯净中产生了静穆的韵律感与节奏感。

在这间没有任何装饰的办公室里，只有极少数的几盏吊灯和照明工具，而阳光、月光、斑斓的树影则纵情地在洁白的墙面地面和顶面上作画，自然将自己的一切都已经镌刻在了这方纯净的天地里。交错的墙面在阻挡、扑捉光线的射入，形成变幻莫测的光影效果。二楼回廊里，天窗射下的阳光在林立的柱体间游走，玻璃墙面将它们收集折射在光滑的墙面和地面上。纯白色的楼梯和整个空间杂糅在一起，来着正是凭借楼板间透出的光影才能找寻登高的路，光明与自由在此化生于白色之间，在心灵的共鸣中带我们进入哲思。 对于设计师而言，白色不仅仅是一种色彩的设计，更是一种设计方法与设计理念。

Designers only a dazzling pure white in the world lying in front of you. Throughout the office space is a white concrete walls and a rectangular column made of a straight line segmentation-based geometry, without an arc shape. Designers with the most abundant and most origin of building elements - walls - the creation of this entire space. The cylinder walls and the surface of tiles and wallpaper and no one, and some walls and even a bit on the lightly coated in white lime cement, masonry the gap is also clearly visible, "count white as black", these columns allow the wall This empty white, in the yes or no, virtual and real, between the Smart manner. Elegant glass facade interspersed the meantime, the vague shapes in the white wall, the sun floating down from top to bottom, produced a solemn and quiet in the pure sense of rhythm and sense of rhythm.

In this there is no decoration of office, only a handful of a few lights chandeliers and lighting tools, and sunlight, moonlight, the shadow is gorgeous heart's content in the white ground and the top surface of the wall painting, nature in all his have been engraved on this side of heaven and earth where pure. Staggered in the block wall, capture the light into the goal, the formation of the vagaries of light and shadow effects. On the second floor corridors, shoot down the sun roof of the cylinder in the inter-lined walk, the glass wall to collect them reflected in the smooth wall and on the ground. Pure white staircase and the entire space hybridity together, to the floor is the virtue of revealing the light and shadow between the climb in order to find the road, light and freedom in this technology was born in a white, between the spiritual resonance of bringing us into overly busy . For the designers, the white color is not just a design is a design method and design.

↑办公室入口 / Office of the entrance.

↑过道 / Corridor
← 二层工作区域 / The second floor large conference room.
↓平面图 / Plan

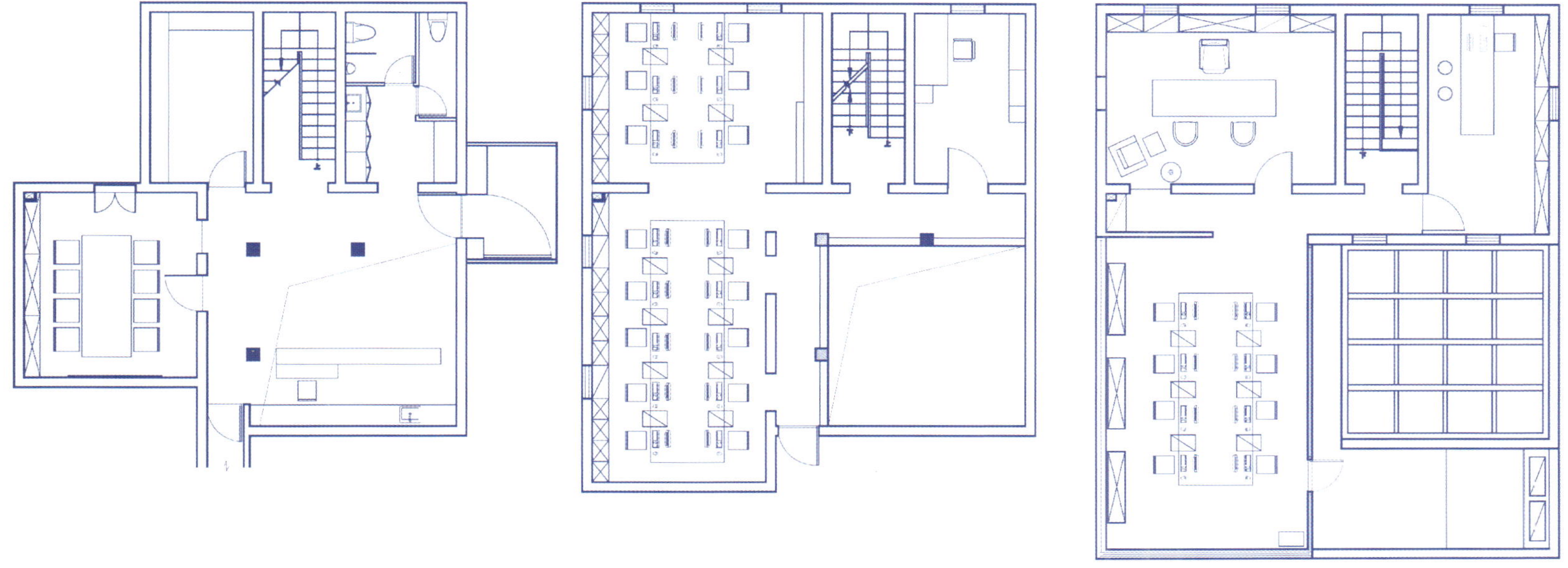

↑总监办公室 / Director of the Office.
→楼梯 / Stairs

↑户外露台 / Outdoor terrace.
← 设计总监办公室 / Design Director's Office.
↓办公区域 / Office space.

KOOLANOO GROUP OFFICE

360圈网站办公室

23

【坐落地点】北京世贸天阶；【面积】1 380 m^2
【设计】杨清华、武雅云；【设计公司】艾迪尔森（北京）建筑装饰工程有限公司
【主要材料】轻质混凝土钙板、麻布贴面、金属吊顶
【摄影】孙翔宇

红色黏土砖墙与黑色金属框，裸露的墙板和管道，到夸张的口号和标语，整个办公空间形式都源于对工业风格的倾慕；从材料的运用、色彩的选择，到家具和灯具的选配，设计师都旨在创造工业感与趣味感并举的效果。

走进大厅，红色黏土配合轻质混凝土钙板的墙面，签字麻布贴面、金属吊顶，直白、刚硬的工业感立刻将你包围。通往办公区的走廊上有红色黏土砖砌成的坐椅，混搭着黑色皮坐垫，粗犷却很实用。走廊墙上的圆形窗户仿佛工厂的排气孔，沟通了办公室与走廊空间，活跃了这并不宽敞的空间。

办公室的设计简洁利落，三面由黑色金属框架的玻璃隔断围合而成，一面是红砖墙，每一间都是如此这般的雷同，好像是一排从流水线上制造出来的标准车间，这是大量生产着奇思妙想和蓬勃激情的车间，门头上的红色报警灯，仿佛随时都在预示工作者的灵感就要迸发，很有一番趣味。最值得一提的是办公室玻璃隔断面上贴的黑白玻璃贴膜，它像涂鸦一样描绘着海底世界的风景。而休息区一改红黑的空间基调，换以沉静优雅的宝蓝色，白色和宝蓝色的桌椅。

Red clay brick walls and black metal frame, wall panels and pipes exposed to the exaggerated slogans and banners, the whole office space forms are derived from the industrial style of admiration; from the use of materials, color choices, to furniture and lamps matching, designers are aimed at creating a sense of simultaneous industrial sense and taste results.

Into the hall, the red clay lightweight concrete with the calcium-board walls, the signature linen veneer, metal ceiling, straightforward, rigid industrial sense immediately you are surrounded by. The corridor leading to the office there is a red clay brick piled chairs, mix and match the black leather cushion, rough very practical. Circular corridor on the wall as if the factory vent windows, communication and corridors of office space, is not active in this spacious room.

Office of the design simple neat, on three sides by a black metal frame enclosure made of glass cut off one hand, a red brick wall, each claimed, are identical, it seems that a row from the line to create standard out of workshops, which is the mass production of the smart ideas and vigorous passion workshops, door head red warning lights, as if all the time indicates that workers should burst of inspiration, something very interesting. Most worth mentioning is that the Office of the glass partition compliment the black and white window film, it is the same as the graffiti depicting the underwater world scenery. The rest area a changed tone red and black of space, change in order to quiet and elegant navy blue, white and navy blue tables and chairs.

↑接待区全景 / Panoramic reception area.

↑通向办公区的走廊 / The corridor leading to office.
↓办公区的隔间 / Office of the compartment.

↑开放式办公区 / Open office area.
↓平面图 / Plan

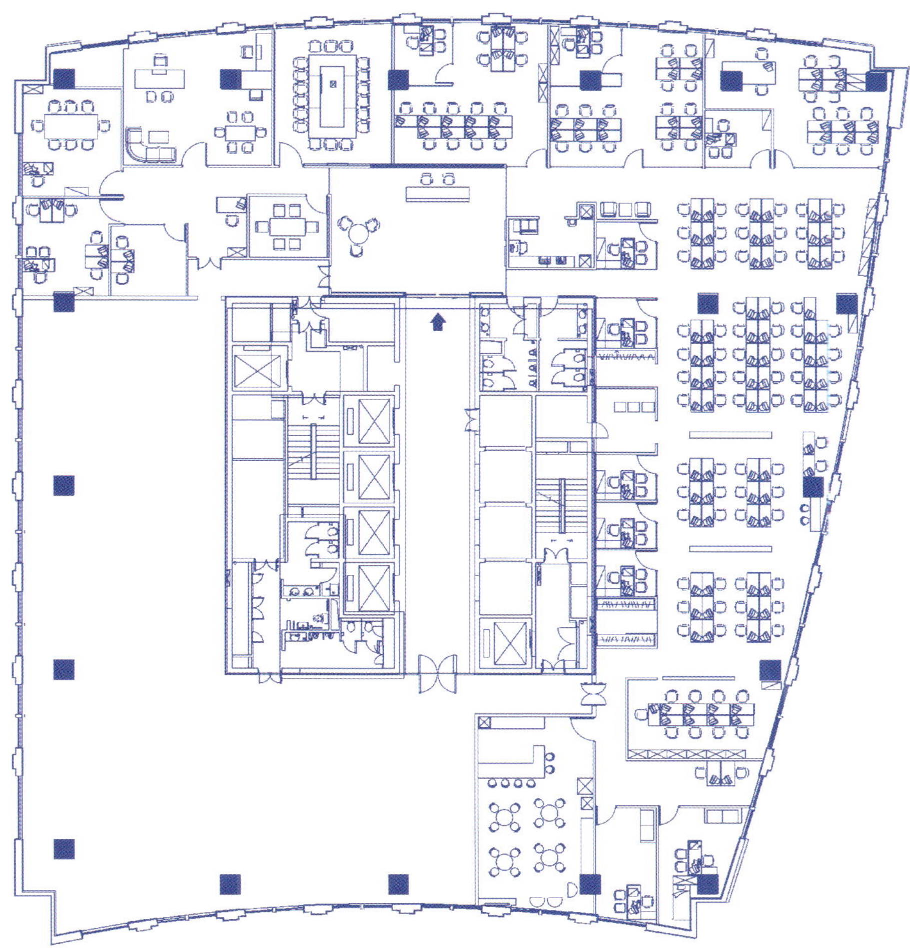

↑ 透过圆窗看开场办公区 / Look through the round window opening office.
← 办公区一角 / Corner office.
↓ 蓝色的空间 / Blue space.

KINGYORKER HEADQUARTERS OFFICE

Kingyorker企业总部大楼

24

【坐落地点】中国台湾台北市内湖行爱路；【面积】3 300 m²
【设计】黄严仕、江姿莹；【设计公司】颉合设计（JAHAA）
【主要材料】胡桃木、玻璃、墨镜、黑云石、镜面不锈钢
【摄影】吴启民、江姿莹

整体总部大楼规划为一整面透明的玻璃量体，尤其在夜晚的时候，室外的黑夜更衬托出建筑的玲珑剔透。简约利落的线条，层次分明的概念，这一设计理念自外立面一直贯彻到室内，延续了建筑上的穿透感与简洁表现。

整体基调以最接近建筑原色的灰色为底，再点缀上白色、黑色、原木色、鲜艳的KINGYORKER企业色——橘色在空间中跳跃。办公区入口，以玻璃及栈板组成一道具科技意象的屏风，以富穿透性的空间元素，界定出内外办公区域，也降低原本量体的压迫感。

5层的策略办公室和6层高层主管室是一整个中通空间，设计师将五层天花楼板退缩进来，刻意表现出挑空夹层设计，将两层变成一层，空间中央以不锈钢及柚木组成的透空阶梯，作为串联的垂直动线，将两层空间联系起来。

7楼附设有宴会厅、酒吧、娱乐间的私人招待所，入内如水池般的圆形等候椅，镜射出户外屋顶花园的倒影。有别于办公区的明亮简敛，设计师汲取东方元素，在空间表现上以暗沉的色彩为基调，从而带给人们一种舒适、安稳的惬意感。

Overall headquarters building planned for an entire volume of a transparent glass body, especially at night time, outdoor night further contrasts exquisitely carved out of the building. Simple, without any lines, the concept of structured, this design concept has been implemented since the interior facade, extending through the building on the sense of performance and simplicity.

The overall tone gray color to the nearest building as a base, then decorated with a white, black, wood color, bright color KINGYORKER companies - orange jump in space. Office entrance, with glass and pallet props composed of an image of the screen technology, the rich penetrating the space element, the definition of inside and outside the office area, also reduced the original amount of the body feeling of constriction.

5-layer strategy for offices and 6-layer top executives in the room is a pass the whole space, designers will come in five-story ceiling, floor contraction, deliberately designed to show the empty sandwich will become a layer of two layers of space central to the composition of stainless steel and teak ladder, as a series of vertical contours, the two-story space link.

7 floor attached to a banquet hall, bar, entertainment, between the private guest houses, entry, such as the pool-like circular waiting chair, mirror reflection shot outdoor roof garden. Different from the office bright Jane convergence, designers learn the oriental element in the performance of the space in order to dull the color of the tone, which gives people a comfortable, cozy feeling secure.

↑外观夜景 / Appearance night.

EXPORT DEPARTMENT

↑玻璃及栈板组成隔屏 / Glass and pallets composed of Partition.
←办公区入口 / Office entrance.
↓ 一层平面图 / 1st floor plan；↓ 二层平面图 / 2nd floor plan.

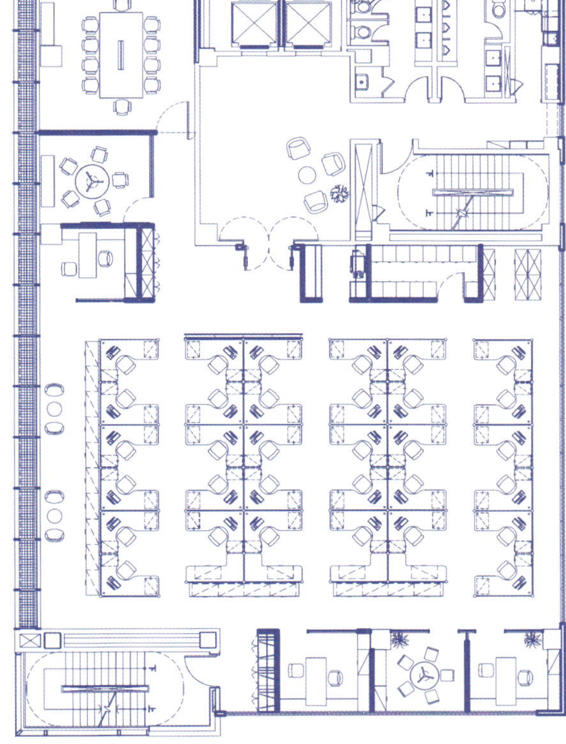

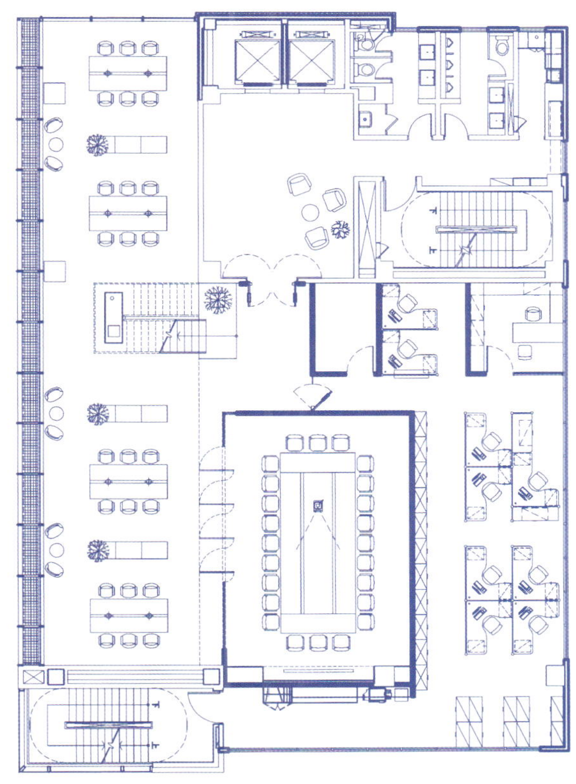

↑透空阶梯 / Permeable ladder.
→空间中央 / Space of the Central.
↓俯视 / Overlooking

↑会议室 / Conference room.
← 楼梯 / Stairs
↓三层平面图 / 3rd floor plan；↓ 四层平面图 / 4th floor plan.

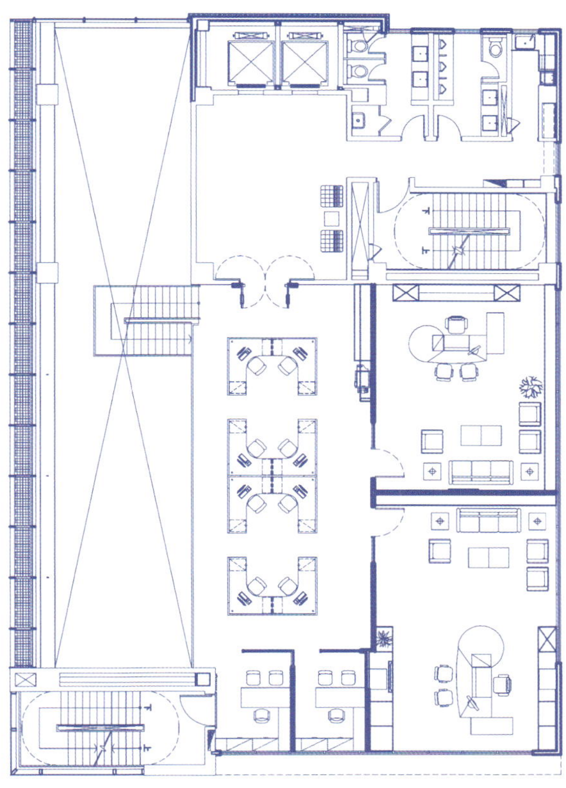

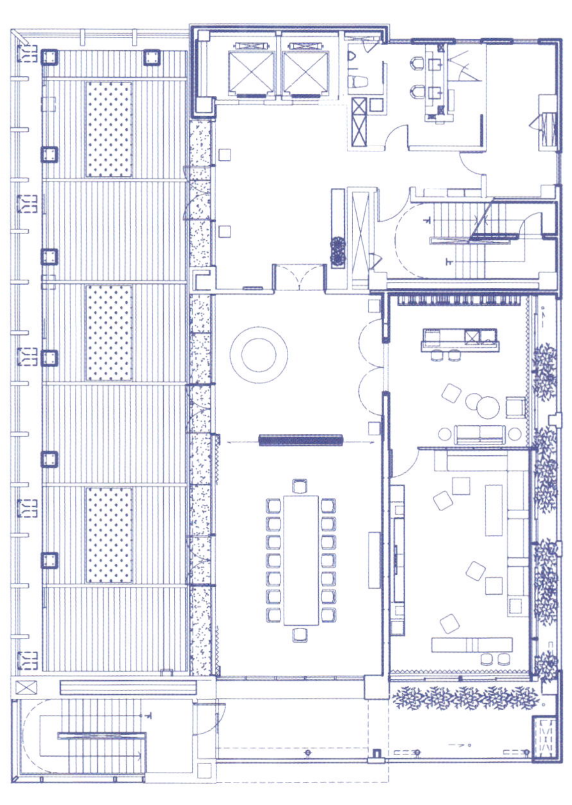

NEW OFFICE BUILDING

深圳市建艺装饰公司新办公楼

25

【坐落地点】深圳市福田区；【面积】1 800 m²

【设计】郭杰

【主要材料】地砖、人造石、GRG、灰玻、木夹板、防火板、灯膜

【摄影】钱翔

本案以“海”作为设计主题，设计师将一个理性、严谨、规矩的传统办公空间转变为了一处开放、清新、时尚，同时又能让人能够沉淀和思考的创意作品。这就是新建的建艺办公楼的审美定位。

采用“海”为主题，以象征“海浪”的白色为主色调，其间搭配以灰金刚材质的地面抽象 “渔网”图案，意用大海的磅礴气势来对抗空间自身的局限。而在整体的结构布局上来看，空间采用了T型台的造型，从电梯间进入，通过地面引导，两边分别进入不同的功能区。右边为总裁办公室、财务室等，开放的办公区则分布在左面。

办公区的设计，亮点放在了天花和玻璃隔墙上。办公区天花的吊顶采用了“海上船帆”的元素，形象地展现了帆船在海浪中的倒影，非常有趣。而在玻璃墙的设计上，设计师跳脱了单纯的隔断作用，在玻璃上设计了一系列动感而“神奇”的线条。无论从哪个角度都可看出不同的视觉效果，既似象征海的波纹，也有象征渔夫洒网的丰收景象。种种精心的设计将海的万千意境与现代办公设计巧妙地结合到一起。

The case with "sea" as a design theme, designers will be a rational, rigorous, rules changes in the traditional office space for an open, fresh, fashion, while precipitation and thinking people can be creative works. This is the new office building construction art aesthetic orientation.

A "sea" as the theme to symbolize the "wave" a white-based color, with gray diamond during the ground material abstraction "fishnet" pattern, meaning to use the momentum to counter the boundless ocean of space limitations. In the overall layout point of view, space station using a T-shape, from the elevator to enter through the ground to guide both sides into the different functional areas, respectively. On the right for the president's office, finance room, open office areas are located on the left.

Office design, bright spots on the ceilings and glass walls on. The ceiling of office ceiling using a "sail the sea" element, image display of the boat in the waves in the reflection, very interesting. In the glass-wall design, designers have deviated from a simple cut off role in the glass to design a series of dynamic and "magic" lines. No matter from which point of view of different visual effects can be seen not only symbolizes the sea, like ripples, but also a symbol of fishermen throwing nets harvest scene. All kinds of elaborate design and artistic conception of the sea of thousands of modern office design cleverly integrated together.

↑过道 / Corridor

↑前台接待区 / Front reception area.
↓前台草图 / Foreground sketch.

↑通向办公区的走廊 / The corridor leading to office.
↓黑白两色是办公区域的主色调 / Black and white is the regional office's main colors.

↑开放式办公区 / Open office area.
← 办公区域 / Office space.
↓过道 / Corridor; ↓天花细部 / Ceiling detail.

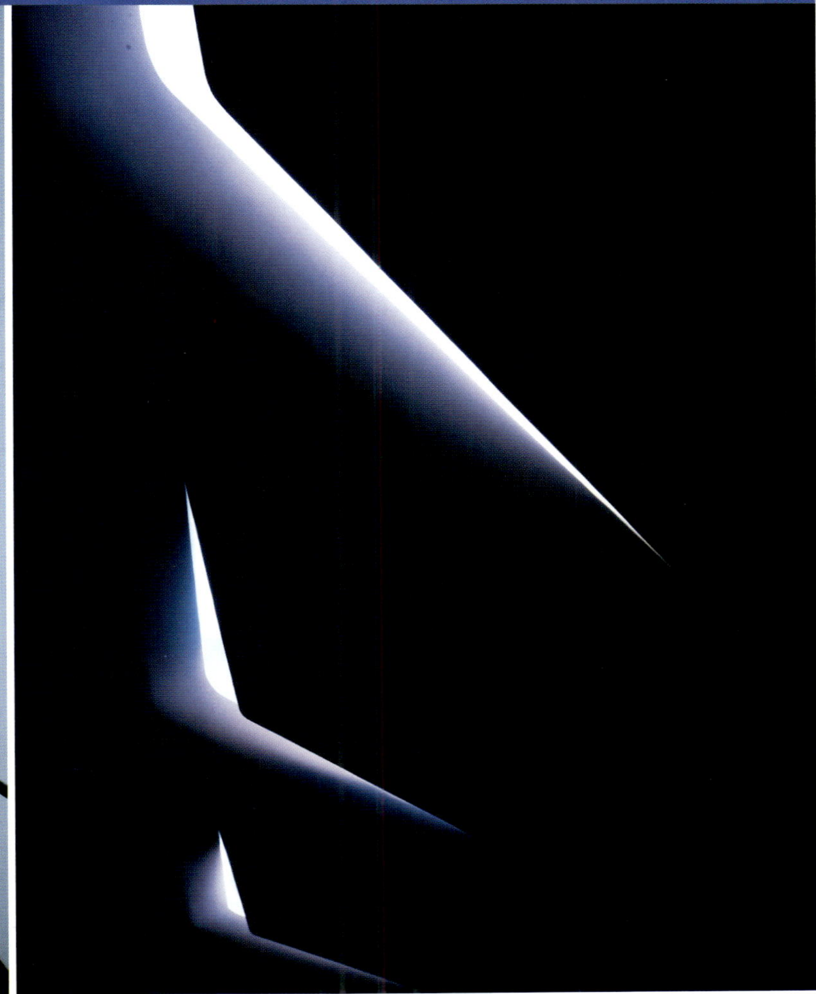

NEW HATCH OFFICE

HATCH办公设计

26

【坐落地点】澳大利亚昆士兰；【面积】10 500 m²
【设计】HASSELL设计顾问有限公司〔澳〕
【主要材料】环保木材、环保橡胶、模块式地毯、玻璃贴膜、胶合板以及亚麻地板
【摄影】Christopher Frederick Jones

本案是国际知名工程顾问公司赫氏（HATCH）的新办公室，HASSELL针对设计任务进行的详细交流是最终获得满意成果的基础，通过现场勘查、当面交流和设计讨论会等形式，对办公楼的功能和设计理念做出明确定位，深切把握赫氏的企业文化和品牌价值。HASSELL在办公空间的设计中对这些设计目标和要求进行了充分考量。

巨大的楼面面积为空间划分提出挑战。HASSELL利用各连接节点作为分区的基点，在尺度巨大的空间内构建社交和工作交流网络。清晰的流线组织图示，渗透到建筑内的日光形成的光影效果，在中心节点周围区域之间喧闹与宁静的对比，形成鲜明的层次感和巧妙的互动。在这些连接节点设置内部挑空和外部楼梯，成为办公空间平面及竖向联系的核心连接节点。

在选材上，历史保护建筑强调对现状建筑形态"轻碰轻触"，不宜大动。选择少数几种饰面，如木材和玻璃，同时适当运用一定的彩色。新工作区和原有建筑元素之间的透明连接使其实现一体化，避免突兀感。对原有木地板进行保留和翻新，并保留周围体现先前居民生活形态的住宅式阳台，与办公空间现在的商务功能形成对照。

This case is an internationally renowned engineering consultancy firm HATCH new office, HASSELL detailed tasks for the design of communication is the ultimate basis of the results satisfactory, through on-site survey, face to face communication and design seminars and other forms of office functions and design philosophy to do a clear positioning, deep grasp the Hershey's corporate culture and brand values. HASSELL in the office space in the design of these design goals and requirements of the full consideration.

Large floor area for space-challenged division. HASSELL use of the connected nodes as the partition point, a huge space in the scale and work to build social networks for the exchange. Streamline the organization a clear icon, penetrating into the buildings of sunlight to form light and shadow effects, in the central node between the regions around the noisy and quiet contrast to the level of a sharp sense of subtle interaction. In these settings to connect nodes pick an empty house and external stairs, into office space plane and the vertical links connected to the core nodes.

In the selection, the historic preservation architectural emphasis on the status of architectural form, "light-touch touch", not making major changes. Choose a few finishes, such as wood and glass, while the proper use of a certain color. New work area and the original architectural elements of a transparent connection between them to achieve integration, to avoid unexpected feeling. Of the original wood floors to retain and refurbish and retain the previous living forms reflect the surrounding residential-style balcony, and office space in contrast to today's business functions.

↑接待区全景 / Panoramic reception area.

↑宽阔的接待空间 / Spacious reception room.
↓公共过道丰富的色彩搭配 / Public hallway with rich colors.

↑接待台后面的小型会客区域 / Front Desk, behind a small sitting area.
↓一层平面图 / 1st floor plan.

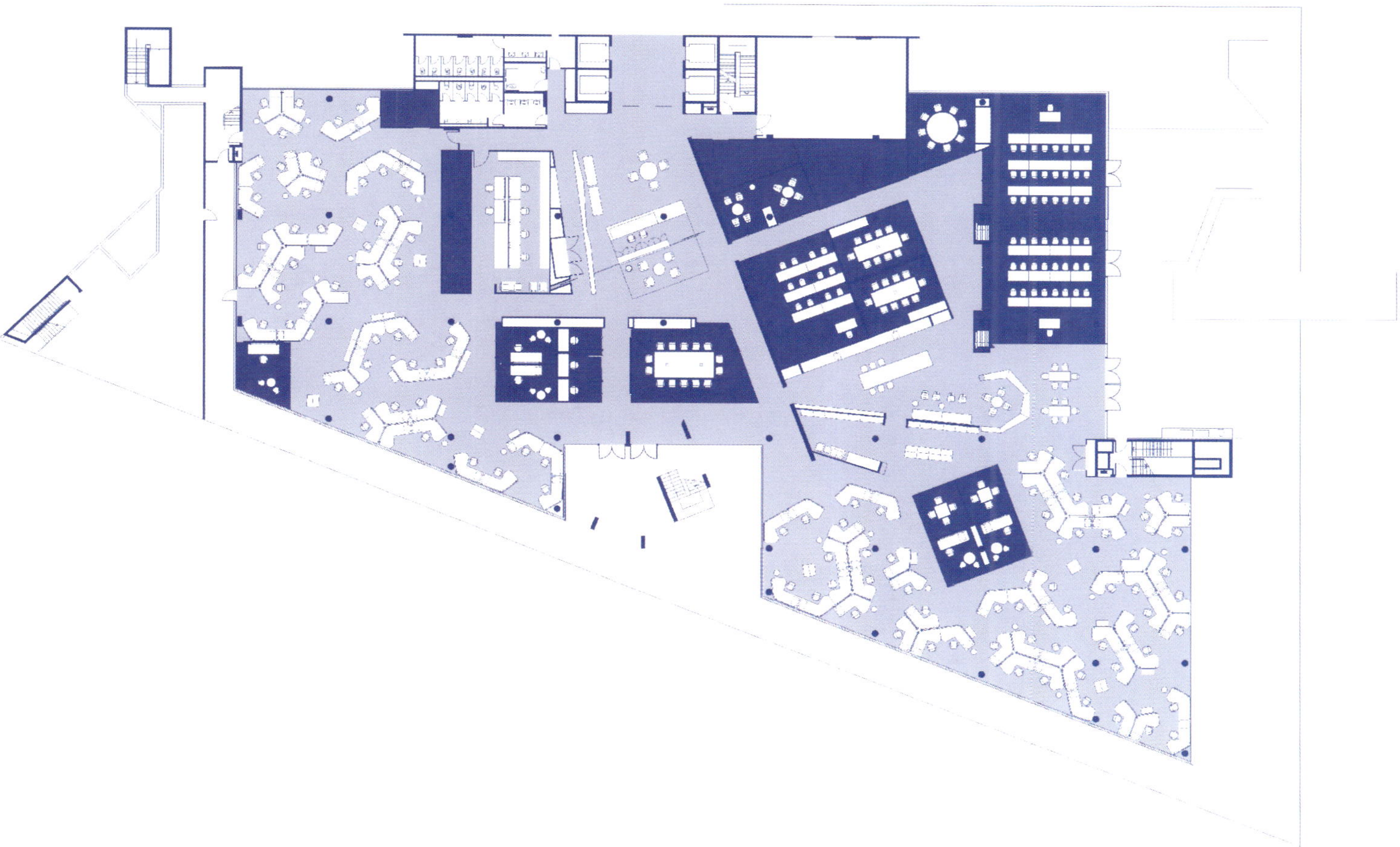

↑透过玻璃看会议室 / To see through the glass conference room.
↓三层平面图 / 3rd floor plan.

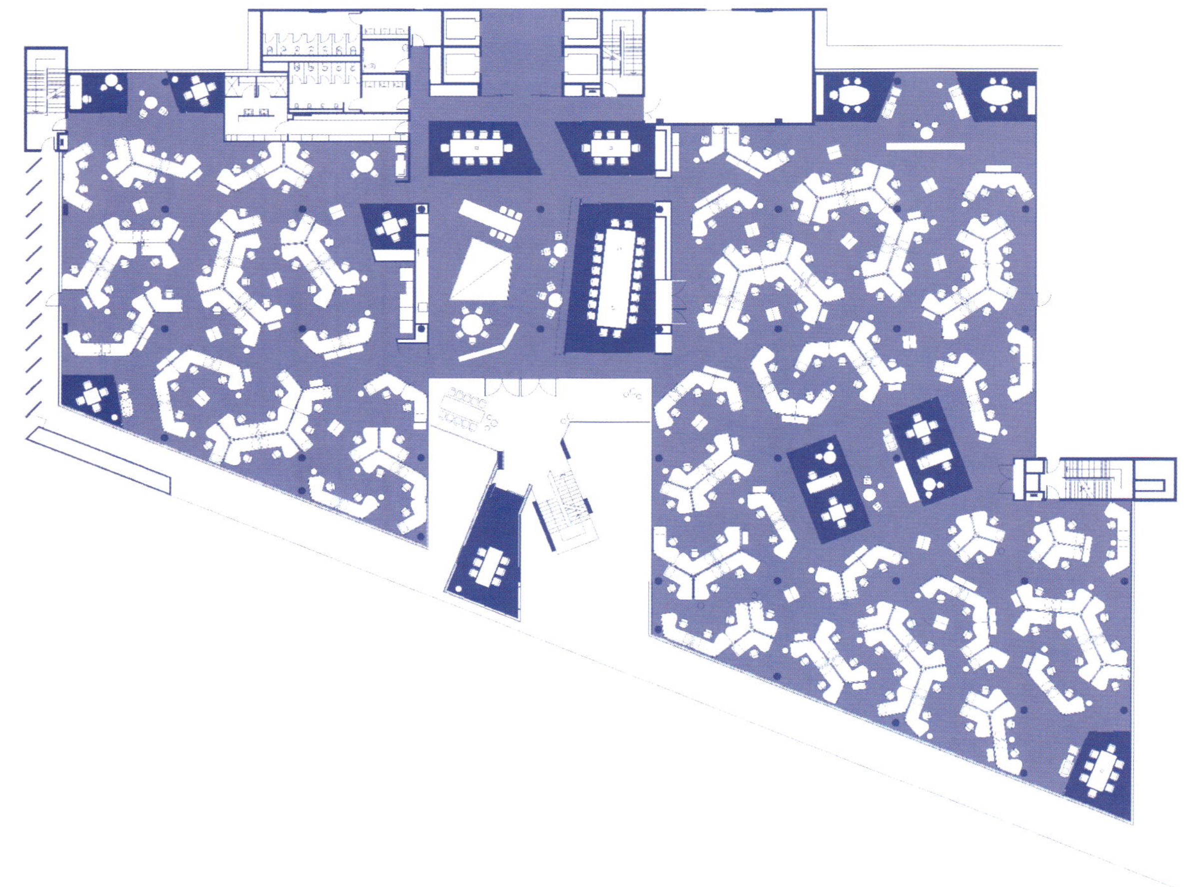

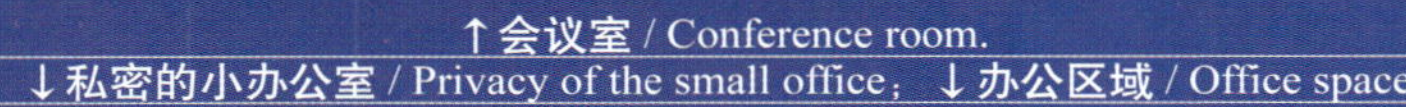
↑会议室 / Conference room.
↓私密的小办公室 / Privacy of the small office；↓办公区域 / Office space.

PLAJER & FRANZ OFFICE

Plajer & Franz办公室设计

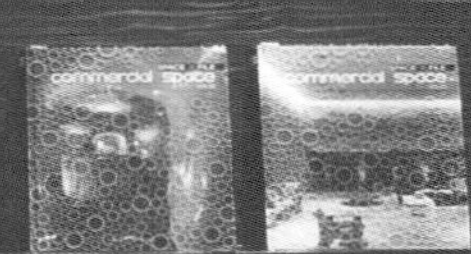

27

【工程名称】Plajer & Franz工作室办公室
【面积】1000 m^2
【设计】Plajer & Franz工作室〔德〕
【摄影】diephotodesigner.de ken schluchtmann

本案为Plajer & Franz工作室45位建筑师、室内与图形设计工作人员的创意基地，工作室的设计师们对这座标志性楼宇敏锐的设计方案，清晰而充分地展现了对细节的钟爱和对款式的认知。

会议室与其他工作区仅以无框玻璃墙隔开，并配以设计独特的门框，使门看起来仿佛在空中飘移，绸缎质感的窗帘可以任意调节空间的隐秘性。从无线局域网到投影卷帘，这些高科技产品都被隐藏在墙壁、屋顶和固定器具中，让新的办公手段与原有的建筑风格很自然地融为一体。

对于客户较少的小型洽谈，执行官办公室里舒适的长沙发是一个再好不过的安静平和的绿洲。沙发、地毯、暖色调主题和装饰性的元素创造出家一般惬意的环境，闪着光的蓝色小鱼游动在嵌入墙体的水族箱中同样也有让人心旷神怡的效果。

工作室的中心是一个开放式厨房，这里也是建筑师们理念付诸生活实践的地方。无论是在吧台，还是沙发，或者在大桌子边，交流和交换灵感始终是最关键的。

The case for the Plajer & Franz Studio 45 architects, interior and graphic design staff in the creative base, studio designers were keen for this landmark building design, clear and fully demonstrated the love of detail and awareness of style.

Conference rooms and other working areas separated only frameless glass wall, and coupled with a unique design of the door frame, so that the door looked as if drifting in the air, satin texture of the curtains can be arbitrarily adjusted the hidden nature of space. From wireless LAN to the projector shutter, high-tech products have been hidden in walls, roofs and fixed appliances, let the new office means and the original architectural style, blend naturally.

For customers with less small-scale negotiations, executive office is a comfortable couch, but no matter how good the quiet and peaceful oasis. Sofas, carpets, warm tones themes and decorative elements of the creation of a monk in general pleasant environment, small fish swimming in the blue flashing light embedded in the wall of the aquarium also is it a relaxing and pleasant effect.

Studio Center is an open kitchen, where an architect, who put ideas into practice areas. Whether in the bar or the sofa, or large table edge, communication and exchange of ideas is always the most critical.

↑开放式厨房 / An open kitchen.

186

↑接待空间 / Reception room.
↓公共过道的色彩搭配 / Public hallway with colors.

↑从开放式厨房望向入口接待区 / From the open kitchen look to the entrance of the reception area.
↓一层平面图 / 1st floor plan.

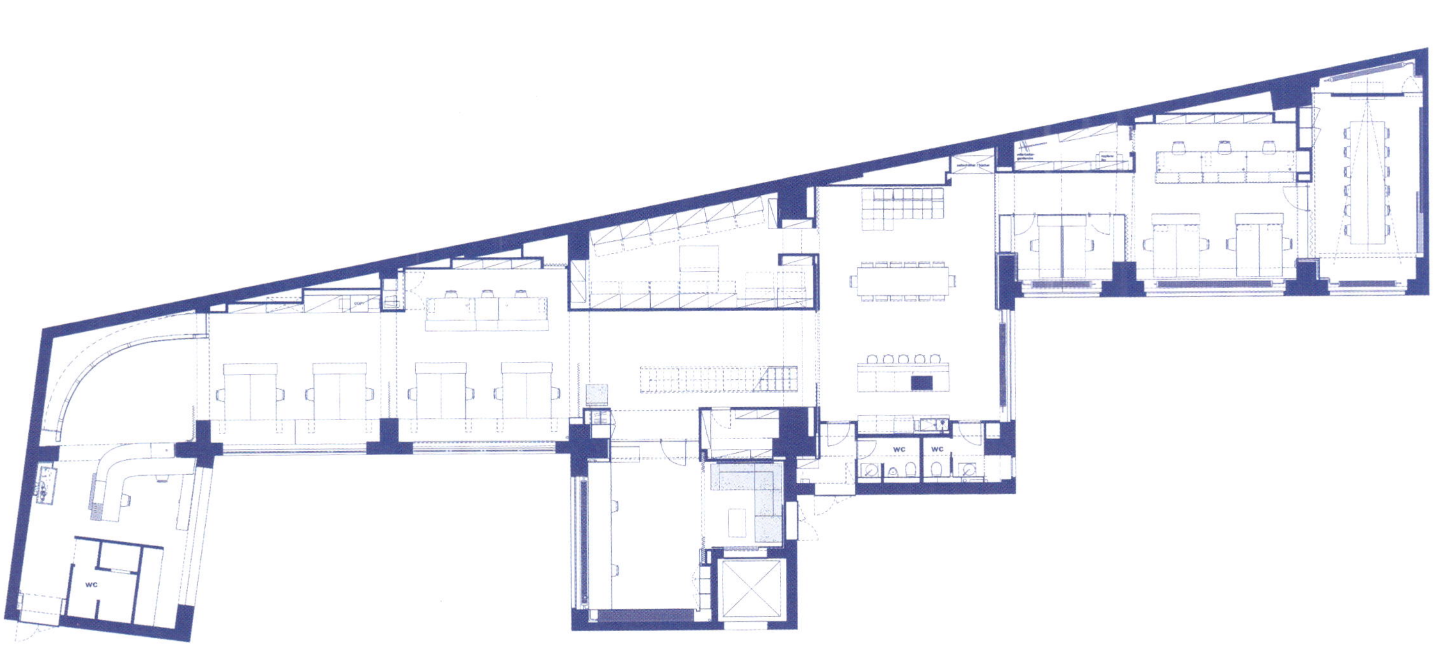

↑一层会议室 / The lower conference room.
← 楼梯设计复杂精细 / Stairs design intricate.
↓一层工作区 / The lower work area.

↑工作区用无框玻璃墙隔开 / Work area with frameless glass walls separated.
↓楼梯口旁以白石子铺就的小空间 / Beside the stairs paved with white stones of the small space.

↑会议室 / Conference room.
↓二层平面图 / 2nd floor plan.

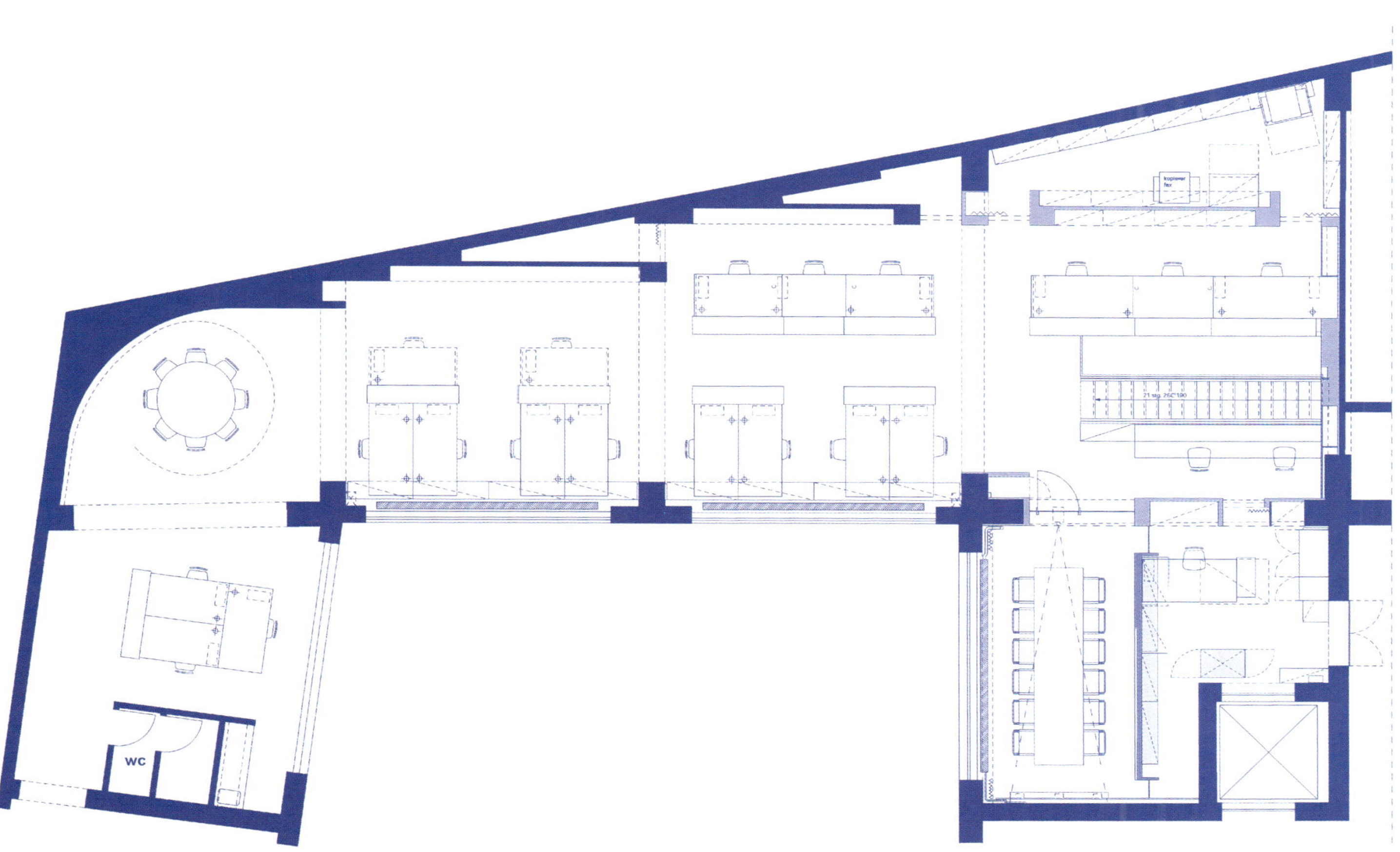

↑执行官办公室 / Executive office.
← 螺旋轨道帘幕营造出的小会议间 / Curtain to create a spiral track of small conference rooms.
↓小型洽谈空间 / Small talk space.

D & G HEADQUARTERS BUILDING

D&G总部大楼

28

【工程名称】D&G总部大楼；【面积】5 000 m^2

【设计】+ARCH Francesco Fresa, Germàn Fuenmayor, Gino Garbellini, Monica Tricario with Miguel Pallarés, Luca Lazzerotti, Fortuna Parente, Magali Roig Liverato

【摄影】Andrea Martiradonna Ruy Teixeira Alberto Piovano

大楼共有5层，其中的一栋还包含了一个两层的地下室。在地上一层，设计师特别规划了一个铺着鹅卵石的漂亮花园。这个绿色的草坪和白色的鹅卵石地面组合而成的花园，给大楼带来了一种特有的清新的美感。

紧邻街道的大楼立面在光线的衬托下剔透晶莹，如百叶窗般半透明的玻璃整齐的排列，就像是大楼的外包装，给人以神圣不可侵犯的感觉。玻璃幕墙的出现可以最大限度地使用自然光源，与此同时它还能够提供一种遮阳的功能，这些玻璃一扇一扇组合起来就像一个过滤板，将外界袭来的强烈的日光过滤成柔和自然的光线投射入进室内。而D&G时尚展厅就设立在这个紧邻街道的位置，来自品牌的色彩斑斓的产品星星点点的点缀着这纯净的立面，犹抱琵琶半遮面的吸引着外界的人们。

休息室在大楼的另一侧，Ron Arad Associates为这里特别设计的家具就摆放在那里，柔和的曲线形设计和大楼直线条的外观形成了一种鲜明的对比。这里的每一处陈设都是经过精心设计和挑选的，在大楼的很多拐角处还设置了等离子电视。直线型的形式和始终如一的自然材质可以在建筑和家具中统统找到。

Building a total of five layers, of which one also contains a two-tier basement. On the ground floor, designers planned a particularly beautiful garden covered with pebbles. The green lawns and white pebbles on the ground of a combination of gardens, to the building has brought a fresh and unique beauty.

Building adjacent to the street facade in the backdrop of light, crystal-clear, such as the shutter-like translucent glass neatly arranged, like a building, packaging, gives a sacred feeling. The emergence of the glass curtain wall for maximum use of natural light, at the same time it can provide a shading functions, which combine a glass door like a filter plate, will hit the outside of the strong sunlight filtering into the soft of natural light into the projection into the room. The D & G Fashion Hall on the establishment in this location close to the street from the colorful brand of products which are scattered dotted with pure facade, still holds partly concealed with the outside world to attract people.

Lounge in the building the other side, Ron Arad Associates is here, specially designed furniture on display, where the gentle curved design and building of the appearance of a straight line forming a sharp contrast. Everywhere where the design and furnishings are carefully selected and many of the corner of the building was also set up a plasma TV. Linear form and consistent use of natural fabrics can be found in both the construction and furniture.

↑开放式的空间设计 / Open space design.

196

↑夜晚的大楼 / Building at night.
↓手绘设计图 / Hand-painted designs.

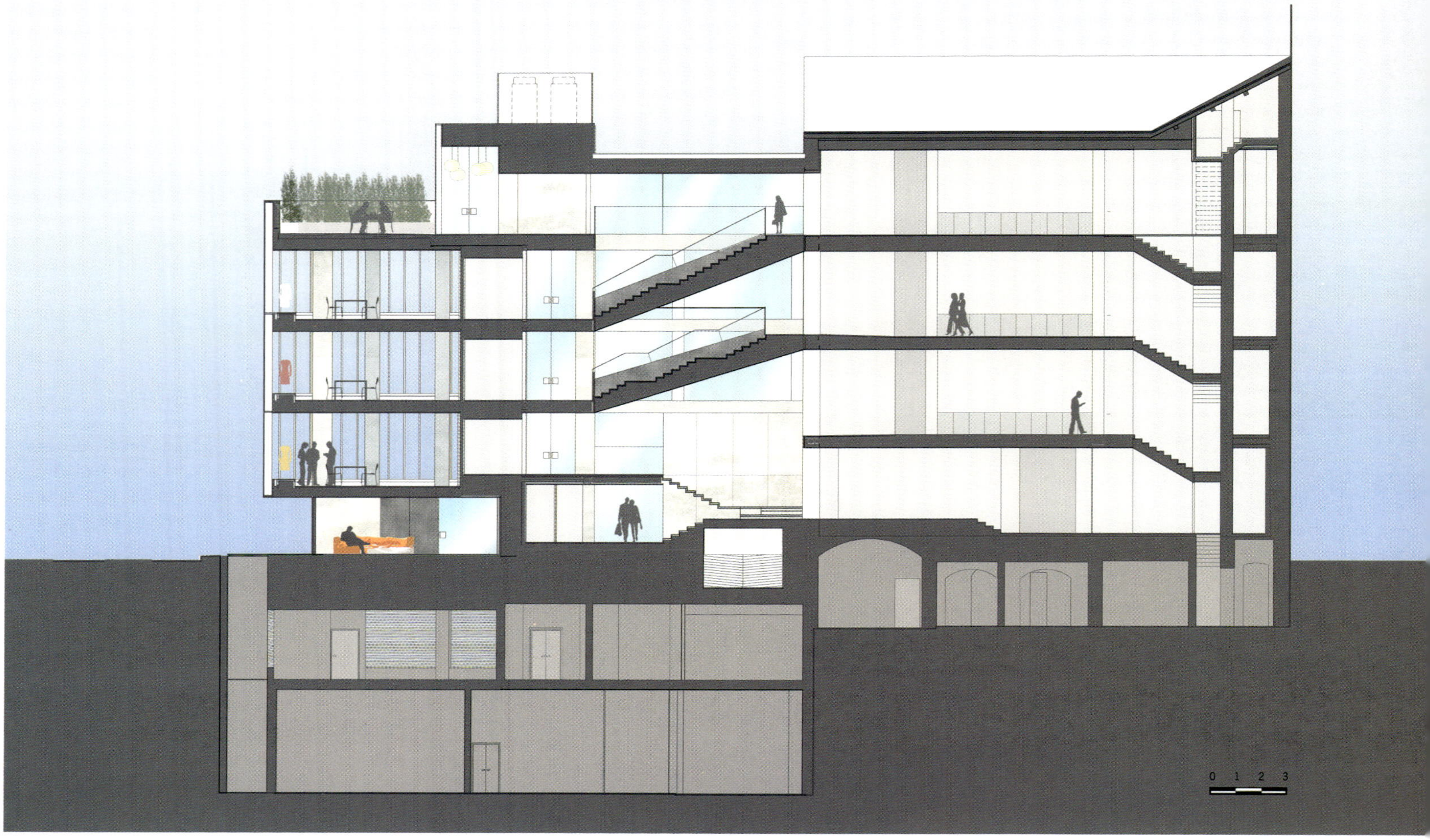

0 1 2 3

↑ 休息室 / Lounge
↓一层平面图 / 1st floor plan.

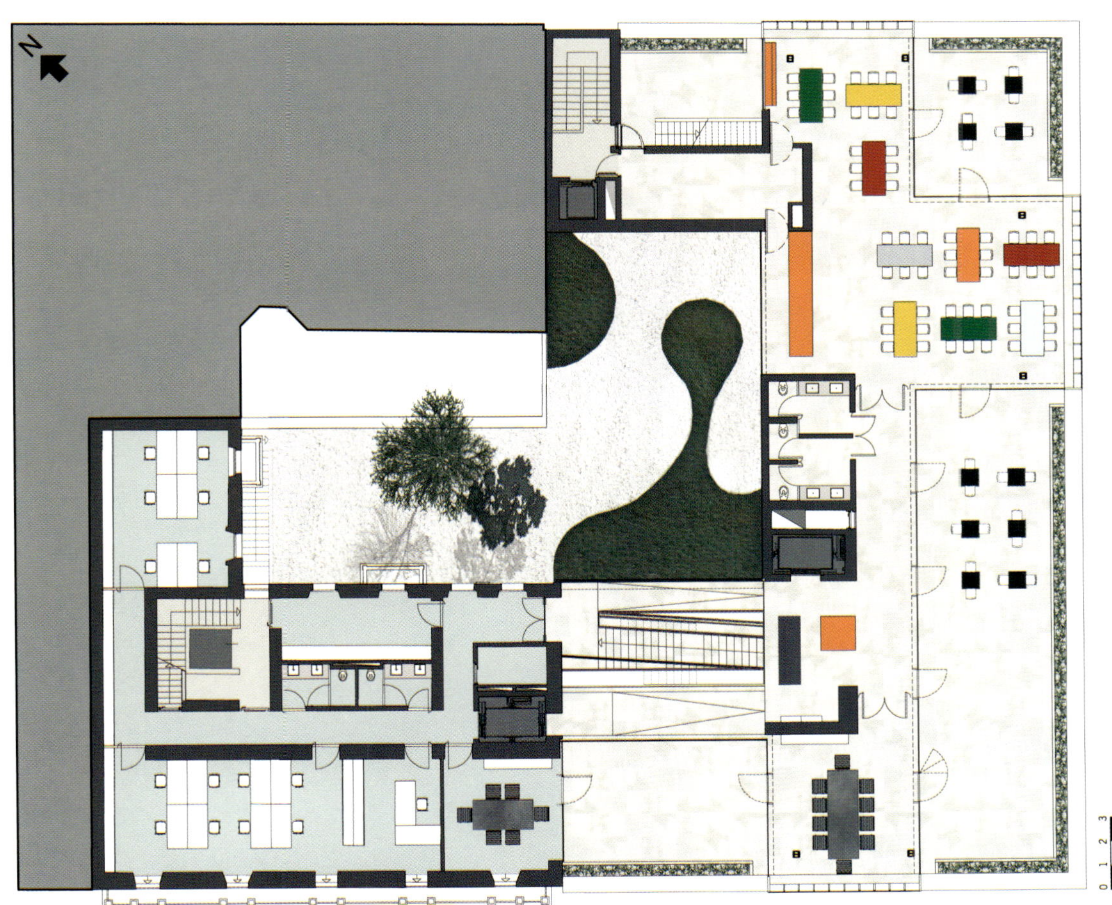

↑内部的结构设计清晰可见 / The internal structural design clearly visible.
← 两栋大楼的连接部分 / Two connected parts of the building.
↓二层平面图 / 2nd floor plan.

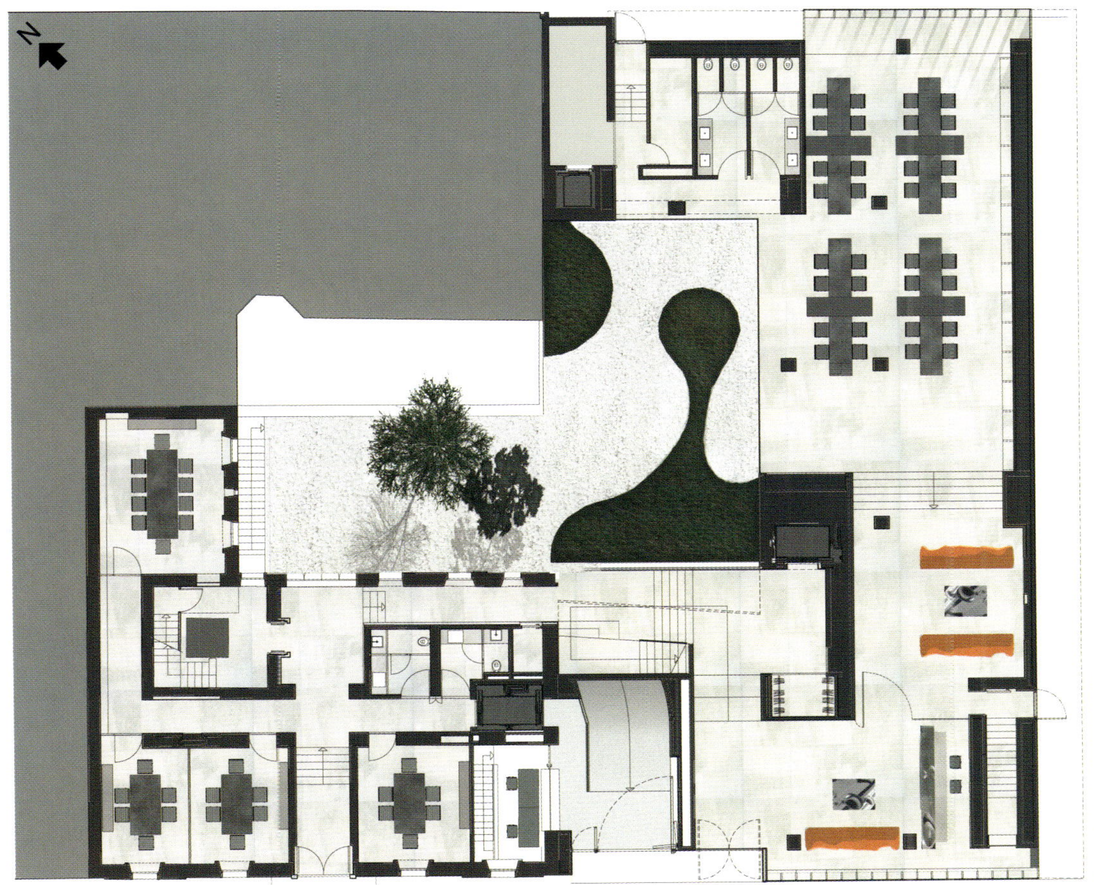

↑开放式的空间设计 / Open space design; ↑楼梯间 / Stairways
← 室内陈设简约优雅 / Simple elegant furnishings.
↓三层平面图 / 3rd floor plan.

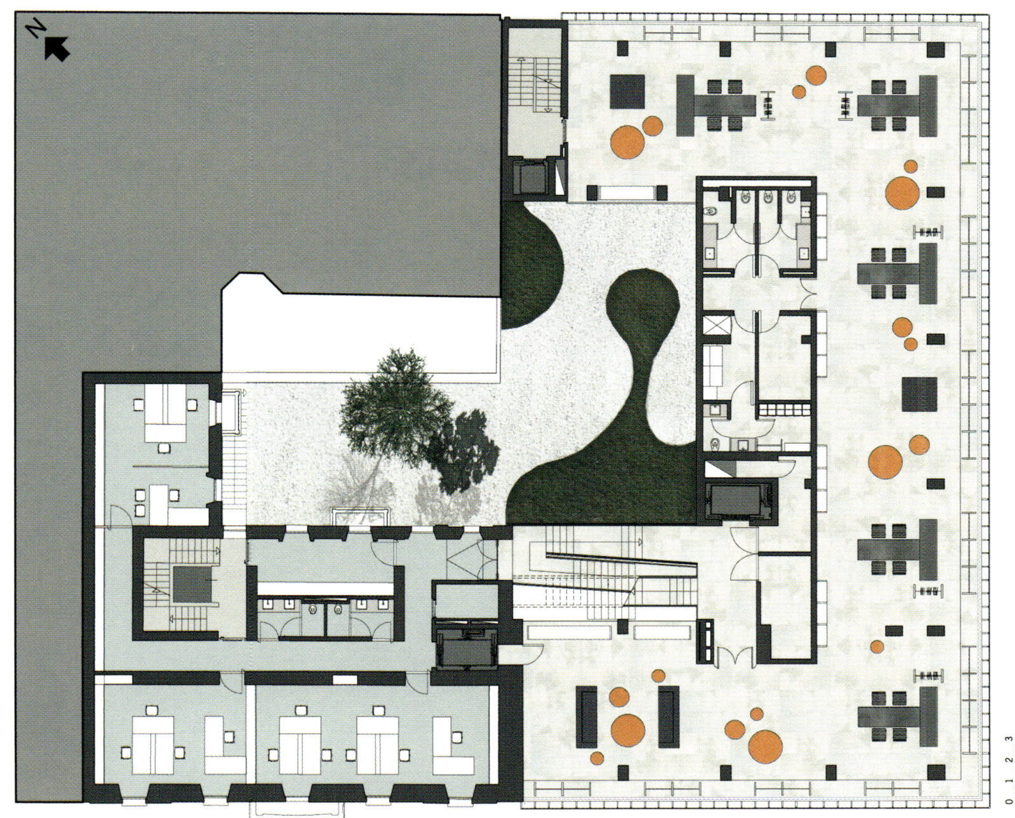

BALS OFFICE

BALS公司办公室

29

【工程名称】BALS公司办公室
【面积】1 820 m^2
【设计】深田恭通〔日〕
【摄影】Nacasa & Partners Inc.

设计师认为BALS公司自己的家具设计就非常有特色，所以他决定就用他们自己的设计产品来装点办公空间。整个方案的风格依然沿用了设计师一贯的低调华丽的感觉，灰色调的紫、华丽的金黄被穿插使用，相得益彰，同时这个空间也在色彩的转换中被巧妙地联系起来。

当从入口进入，你便会一眼看到那些漂亮的当代艺术作品安然地悬挂在白色的墙面上。大堂的地面使用了"人"字形紫色胡桃木材质的地板铺设而成的，它们也是特别订制的。而室内随处可见的青铜色镜面制作而成的墙面被大面积的使用，为空间奠定了华丽的基调，并且增加了这里虚拟空间的面积。在员工休息室，长桌两旁特意摆放了渐变色彩的椅子，它们的色彩和墙面、地板形成了一种强烈的对比，并且让这里多了一份视觉上的节奏感。

主席办公室的特色也力求在符合功能性的同时体现出家具设计的原始美感，简约优雅的气氛伴随着柔和的灯光，庄严之余不失艺术格调。会议室成功地塑造了一种稳重、温馨的氛围，简洁的灯光打在墙板上，反射着柔和的光线，走廊的隔墙用的也是青铜色的玻璃制作，小会议室的那一面，柔和的灯光可以轻易地透过玻璃墙散落下来，柔化空间的同时，也释放了工作空间的压迫感。

Designers that BALS the company's own furniture design is very distinctive, so he decided to use their own design products to decorate the office space. Program as a whole follows the designer's style has always been low-key feel gorgeous, gray tone purple, gorgeous golden yellow are interspersed use, complement each other, while the color space conversion are also being subtly linked.

When entering from the entrance, you will see one of those beautiful works of contemporary art hung peacefully on the white wall. The lobby on the ground using the "human"-shaped purple walnut flooring material made of, they are also special ordered. The interior can be seen everywhere in the walls made of bronze mirrors have been large-scale use of the space and laid a gorgeous tone, and increase the area of virtual space here. In the staff lounge, a long table gradient color on both sides of specially placed chairs, their color and walls, the floor formed a strong contrast, and so here than in a visual sense of rhythm.

Office of the President also sought to meet the performance characteristics of furniture design at the same time reflect the original beauty, simplicity and elegant atmosphere, accompanied by soft lighting, solemn in style with the art. Succeeded in shaping the conference room of a stable, warm atmosphere, simple lighting hit the wall, the reflection with soft light, the corridor walls are also used bronze glass production, small conference room of that side, soft The lights can be easily scattered through the glass wall down and softens the space, also pressed for the release of the work space.

↑入口 / Entrance

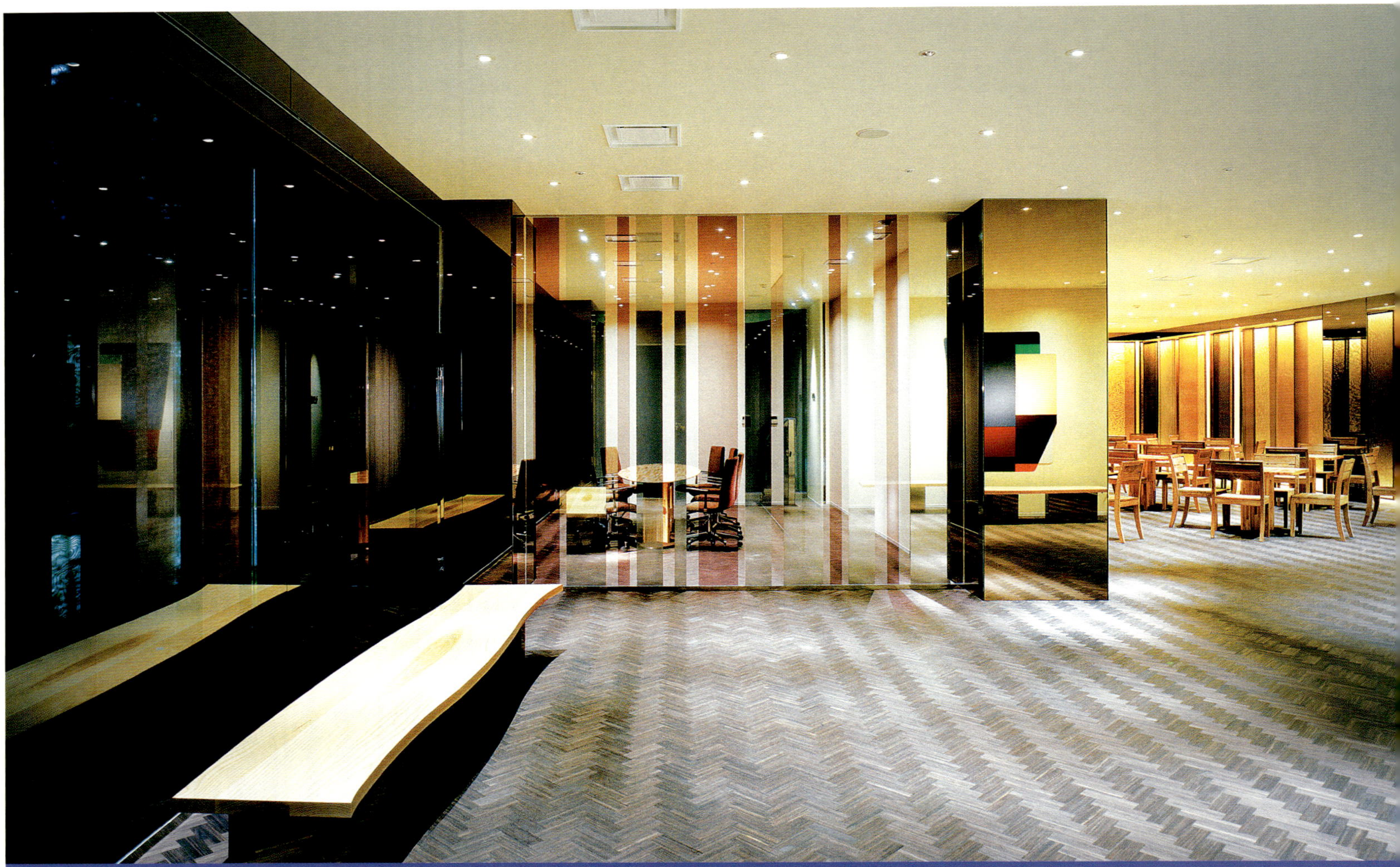

↑玻璃、木材与金属三种材质的运用 / Glass, wood and metal three kinds of materials should be employed.
↓休闲放松的环境 / Casual relaxed environment.

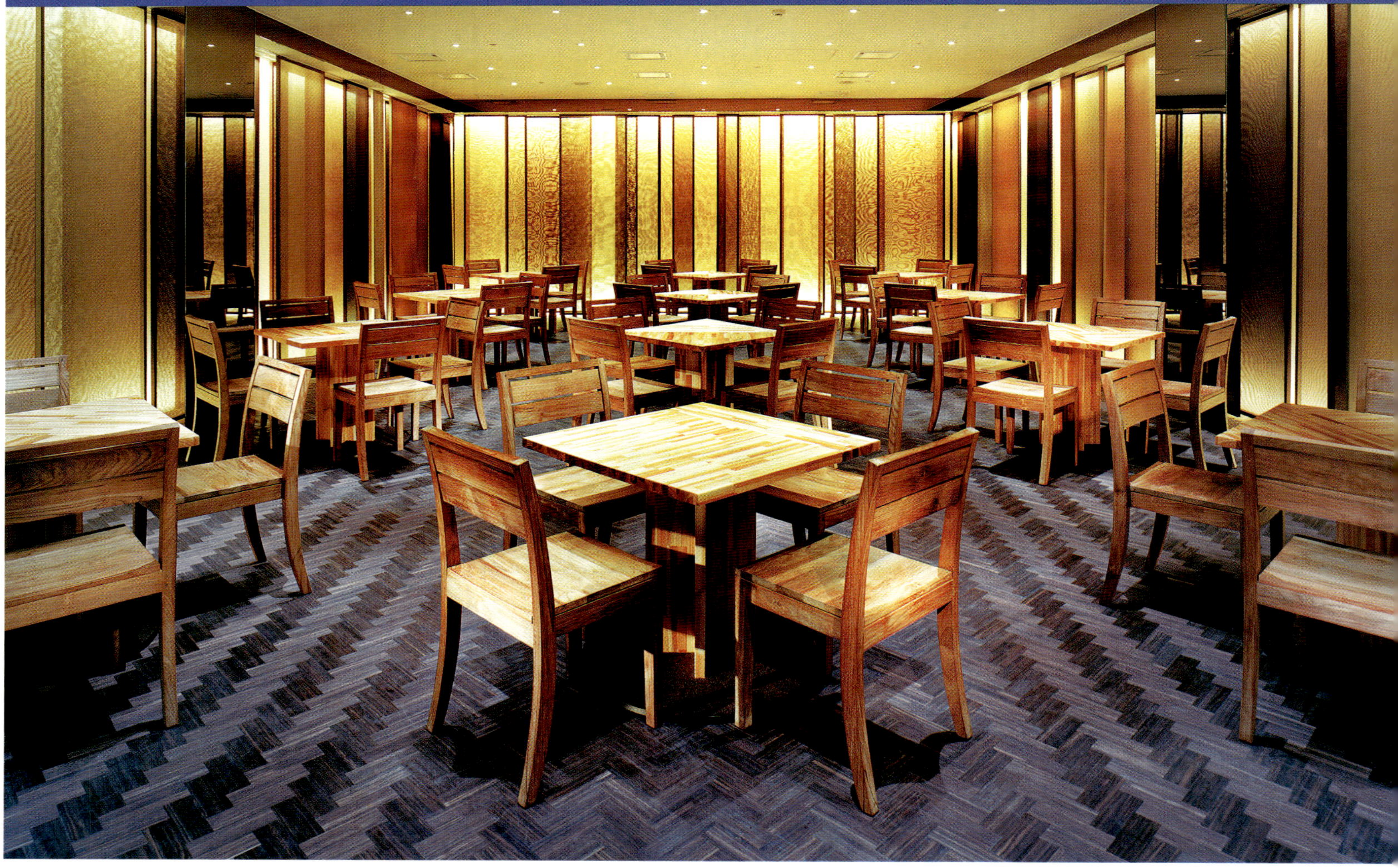

↑ 休息室 / Lounge
↓ 平面图 / Plan.

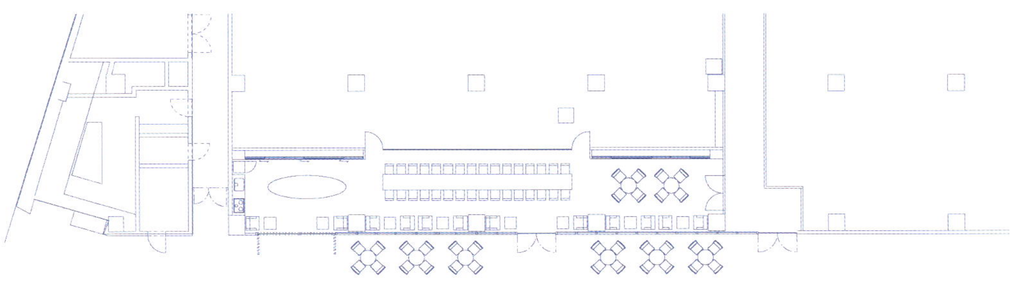

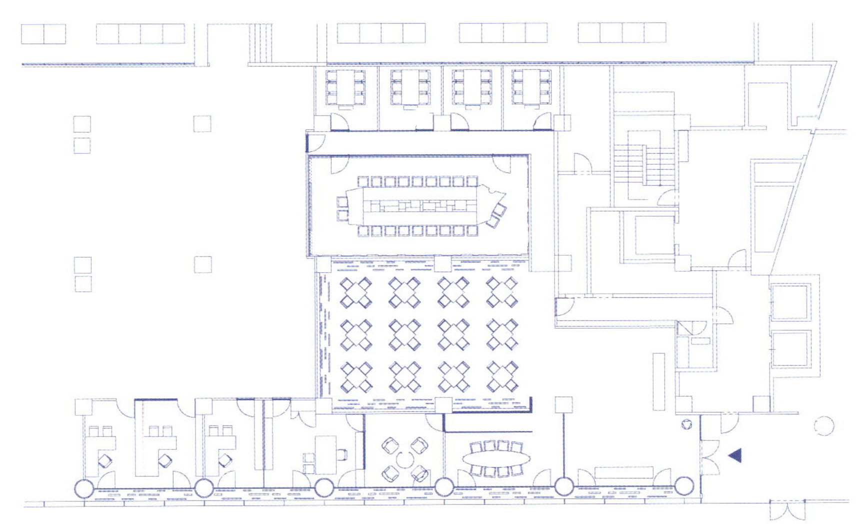

↑小会议室 / Small conference rooms.
↓主席办公室 / Office of the President.

↑ 员工休息室 / Employee lounge.
↓ 会议室 / Conference room.

STU

PURE SOHO

纯粹SOHO

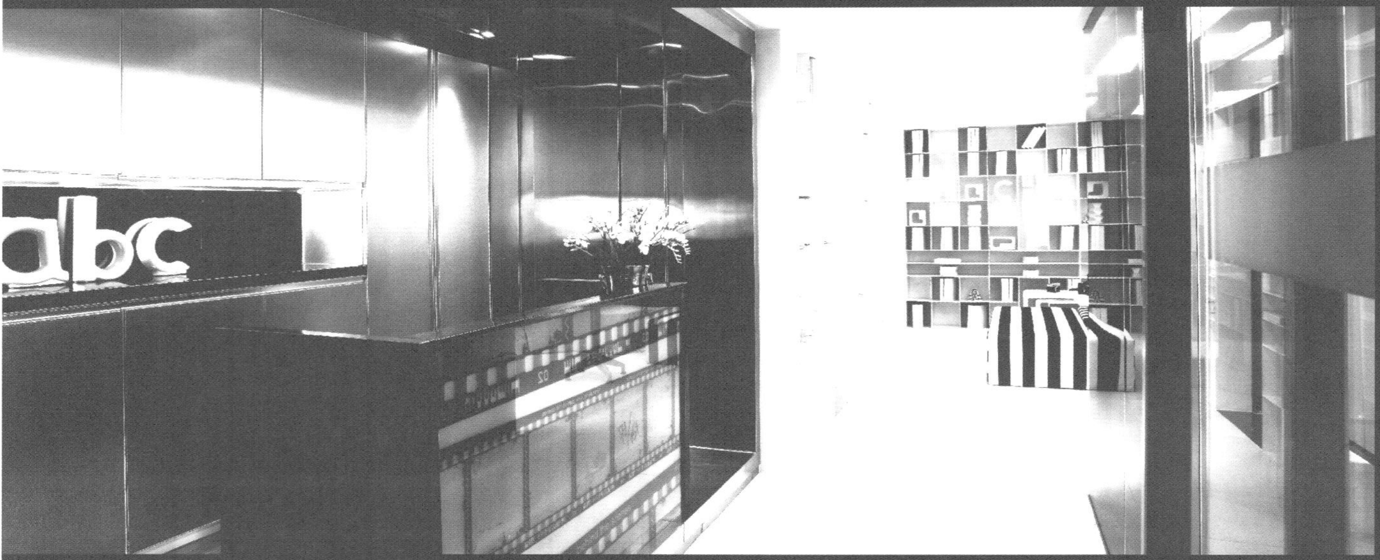

30

【坐落地点】广东省深圳市
【面积】60 m^2
【设计】李益中
【主要用材】地砖、白色氟碳漆、不锈钢、铝板

这个面积仅60m^2的户型，原始平面及为简易，正适合于打造成一个SOHO空间。

作为一个SOHO空间，书籍，文件的储存必不可少，因此设计师巧妙的将墙设计书架形式，既将工作区同洗手间，衣帽间分隔开来，也保持了视觉上的完整性，同时也起到收纳储藏的作用。客厅设置成多功能区，摆放组合家私，可以根据组合家私不同的摆放方式设置成工作区/会客区/影视区的模式，甚至可以进行小型的PARTY，空间得以充分利用。

总的来说，这是一个干净，纯粹，富有多变性及组合性的功能空间。

An area of 60 square meters of Unit, the original plane and for simplicity, are suitable for playing create a SOHO space.

As a SOHO space, books, file storage are essential, so designers will be cleverly designed wall shelves form, both the working area with toilets, cloakroom separated, but also to maintain the integrity of the visual, but also play a incorporating storage role. Living room set into a multi-zone mix placed furniture can be placed under the portfolio of furniture of different ways to set into the work area / reception area / video area of the model, even for small-scale PARTY, space can be fully utilized.

Overall, this is a clean, pure, rich variability and composition of function space.

↑特色书柜 / Features bookcase.

↑入口接待区 / Entrance reception area.
← 工作区 / Work area.
↓平面图 / Plan

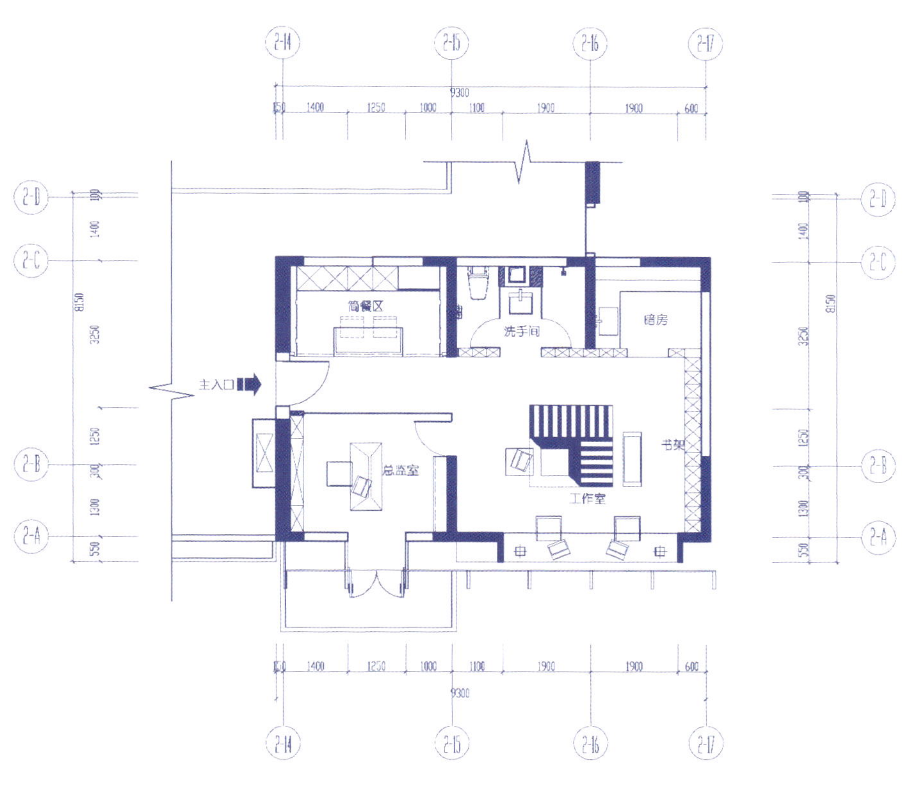

平面布置图 COLLOCATION PLAN
SCALE: 1/75

↑工作室一角 / Studio corner.
↓工作室全貌 / Studio picture.

↑SOHO办公室一角 / SOHO office of the corner.
↓SOHO办公室 / SOHO office.

FOLDING THE SPACES

台北知了工作室

31

【坐落地点】中国台湾台北市
【面积】70 m^2
【设计】陆希杰；【参与设计】林欣怡、潘品臻、吕雅惠
【摄影】Marc Gerritsen

知了工作室虽然面积不大，但是在对空间创新性的规划后却并不显得局促。大面积的窗户让光照亮室内，天花板则保留了原建筑物裸露的管线，整个空间舍去传统的隔间墙面，取而代之是滑动式黑铁金属拉门，配合着自然的光线及白色基调，让空间显得非常宽敞。黑铁拉门在闭与开之间，造就了空间的弹性使用及工作室的趣味性，依照不同的使用功能或不同的使用者，呈现出与隐私性的会议室或是开放式的画廊或摄影展等不同的样貌。可以说，知了工作室在保留艺术性的同时也兼顾了自然、人文及创意于一体。

在空间功能习惯被看做建筑材料后，生活方式也往往会随着空间造型的改变而改变。所以，空间不管是高低还是错落，不论是凹凸还是平坦，都是房子功能的一部分。知了工作室的空间主要分为导演办公室、会客室、道具间、试片室，以及拍片前的试镜场所与剪辑室等，这些空间通过黑铁拉门时而相连时而分割，显示出一种灵活搭配的变化。而空间内搭配独特的家具与大面积的书柜，以及造型独特的顶灯和装饰物，都更显示出一种人文气息的特质。

Cicada studio though small in size, but in the space planning of innovative but do not seem cramped. Large windows allow light lit room, the ceiling is to retain the original building exposed pipeline, the entire space rounding traditional compartment walls, replaced by a sliding Iron metal sliding doors, combined with the natural light and white tone, so that it is very spacious room. Iron between the sliding door in the closed and open, creating the flexibility of space use and fun studio, in accordance with the use of different functions or different users, showing a conference room with privacy, or open gallery, or photographic exhibitions of different appearance. Can be said that cicada studio art at the same time to retain both the natural, cultural and creative in one.

Accustomed to being in space capabilities as building materials, the way of life tend to change with the change in the shape of space. Therefore, space, whether high or low, or scattered, whether it is convex-concave or flat, are part of the house functions. Cicada studio director of space can be divided into offices, meeting rooms, props, the specimen room, as well as the film before the audition venues and editing room and so on, and sometimes these spaces connected by sliding doors and sometimes Iron division, shows a flexible with changes. The space with unique furniture and large area of the bookcase, as well as exotic dome lights, and ornaments, all showed the characteristics of a humanistic atmosphere.

↑开放的办公区域 / Open office space.

↑分割巧妙的空间 / Split clever space.
↓可拉伸的铁门将空间一分为二 / The gates of space can be stretched into two.

↑铁门滑开的会议室 / Sliding iron gate open conference room.
↓平面图 / Plan

↑颇具人文气息的办公环境 / Rather civilized atmosphere of the office environment.
←↓可拉伸的黑色铁门 / Can stretch the black iron gate.

↑另一个角度的会议室 / Another perspective of the conference room.
↓由黑色铁门围成的会议室 / Surrounded by a black iron gate of the conference room.

↑特色顶灯及装饰物 / Features dome light and decorative items.
↓会客室 / Reception room.

SOHO SHOW FLAT

SOHO样板间

32

【坐落地点】北京三里屯工体北路南侧
【设计】隈研吾〔日〕
【参与设计】ASANO
【摄影】贾方

设计强调体验性和现象性，以取代形态性的直白。自由弧线的建筑外皮下包裹的是一个能够感受到“自由”在穿行的场所，这种自由的感觉可以帮助你找寻以往的记忆。砖元素的形状、色彩还有那些活泼的透过树叶映在其上样板间一层的吊顶、内墙面通过大小不等的弧线机理给我们再现线性空间里特有的流动感觉。水中涟漪般的不规则图案交织叠合创造出一个纯净的三维空间。地面的流光透过柔和的磨砂玻璃，晶莹透亮。

金属材料的绞线并没有给人生冷僵硬的感觉。二层的商务空间极具现代气息。白墙面上镶嵌着自由排列的金属书架，外形是扁平的平行四边形，整体呈现出跳跃流动的感觉。长弧线的会议桌上配以细长的深色灯具，为整个袋形空间增添时尚的活泼气息。

居住样板间被安排在三层和四层，一切皆是素面登场，清新淡雅的风格极具东方魅力。顶面和墙面以白色调为主，家具的外形简约方正，质感宜人。与弧形外窗形成刚柔对比。纯净的室内空间隐含的是精巧的制作技术，块面之间的干净交接考验着设计师的构造逻辑。

Designed to emphasize experience and phenomena of nature to replace the form of straightforward. Free arc under the skin of the building is wrapped in a feel the "freedom" in the walk through the place, this feeling of freedom can help you find the memory of the past. Brick element shape, color and lively through the leaves there are those on which model reflected in the inter-floor ceiling, inner walls of the arc through the mechanism of different sizes to our unique reproduction of a linear space, the flow of feeling. Ripples in the water like an irregular pattern superimposed woven to create a pure three-dimensional space. Streamer through the soft ground, frosted glass, translucent crystal.

Stranded metal materials does not give the feeling of cold rigid. Two-story commercial space for a very modern flavor. White walls inlaid with free arrangement of the surface of metal shelves, the shape of a flat parallelogram, the overall flow showing a jump feeling. Accompanied by a long arc of the slender, dark conference table lamps, for the whole bag-shaped space to add a stylish and lively atmosphere.

Living model that they would be scheduled for three and four, all are is a prime face the stage, fresh and elegant style of great oriental charm. The top surface and walls with white tone-based, furniture Founder simple shape, texture and pleasant. With the arc outside the window of the formation of rigid-flexible contrast. Pure interior space is implicit sophisticated production technology, clean transition between the block face the test of a designer's structural logic.

↑建筑表皮 / Structure's facade.

↑↓自由弧线的建筑外皮下包裹的是一个能够感受到“自由”在穿行的场所，这种自由的感觉可以帮助你找寻以往的记忆 /
Free arc under the skin of the building is wrapped in a feel the "freedom" in the walk through the place.

↑↓样板间一层的吊顶、内墙面通过大小不等的弧线机理营造出一种线性空间里特有的流动感 /
Model between the layer of ceiling, inner walls of the arc through the mechanism of different sizes to create a linear space, unique sense of movement.

↑另一个角度 / Another perspective.
→办公室 / Office
↓楼梯 / Stairs

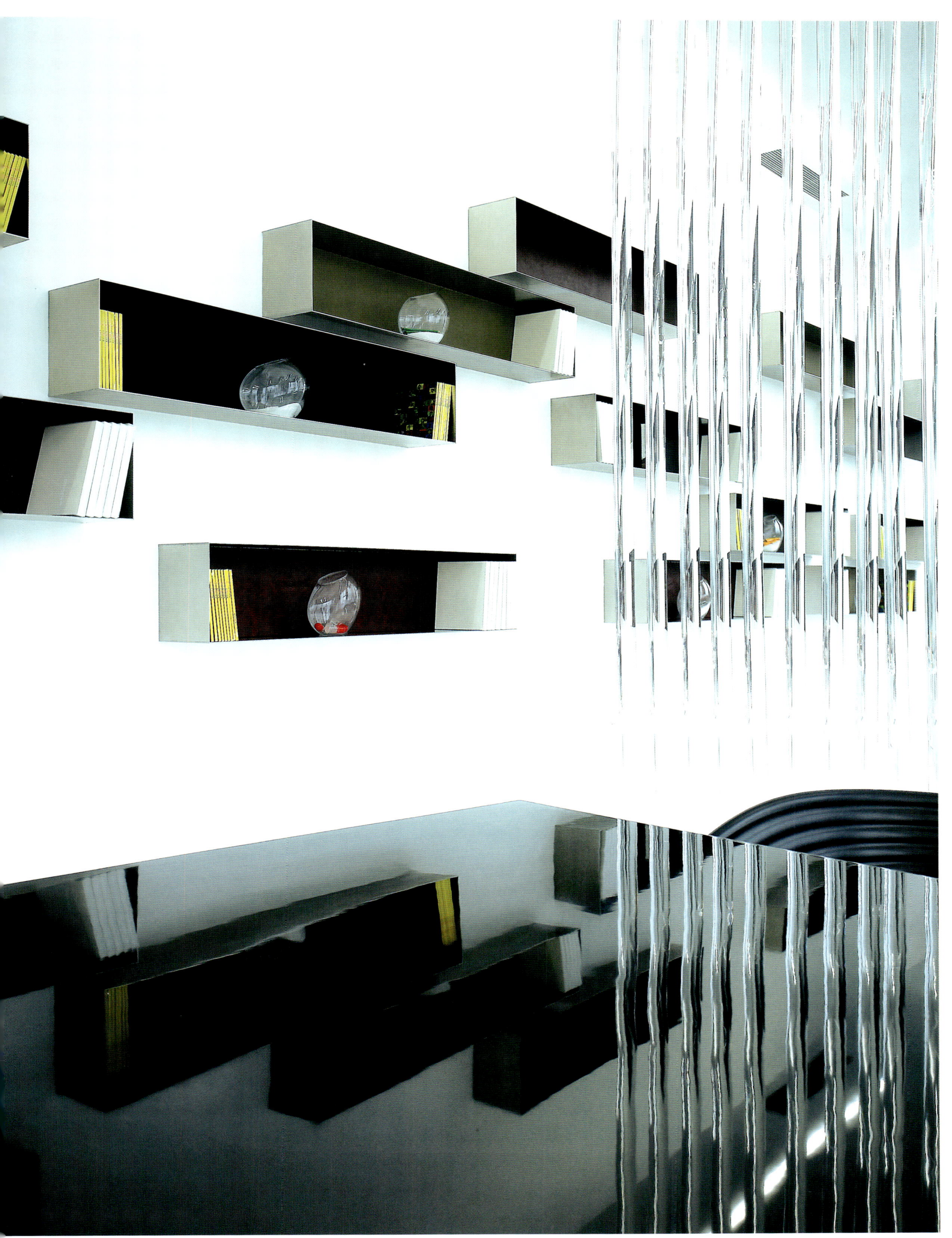

↑办公室 / Features dome light and decorative items.
←走廊 / Corridor
↓会客室 / Reception room.

ANNEX STUDIO

Annex工作室

33

【坐落地点】美国加州洛杉矶；【面积】1 068 m²
【设计】克莱夫·威尔金森建筑师事务所〔美〕
【参与设计】克莱夫·威尔金森、亚历克西斯·拉巴波特
【摄影】班尼·陈摄影工作室

原有建筑与新设施完全开放，共同改善了这一建筑。主要的设计充分利用了原银行大厅的高度，包括一系列夹层（作为团队工作区），穿插了一些临时桌椅，这样既可满足手工工作的需要，又能在此进行计算机辅助设计。

这些空间包括"Tank"、"Pool"、"Wave"。其中，"Tank"是一个定制的蓝色玻璃箱，放在嵌入的柱子上。一条充满液体的"充填输送管"，形成蓝色的氖灯管，缠绕在玻璃箱的四周，为下面的蓝色玻璃墙增添梦幻光彩。"Pool"是一个抬高的木质平台，平台中央有一个大的凹陷，装饰着游泳垫。这一空间是团队进行非正式讨论的地方。旁边定制的休息椅有可调节的扶手，为带手提电脑来此工作的人提供了一个放松的，类似游泳池边的休闲椅。"Wave"是一个独立的夹层，看起来像是浮在空中。一张大布告板和方桌形成一个空间，学生即可在此进行个人创作又可相互合作。

工作室的墙面由印有仙人球图案的Marimekko布装饰板包裹，这样可以减弱环境噪音。在一层，定制的低矮榻榻米台子上铺着人造毛地垫，这为非正式工作的环境增添了多样性。照明设备能照亮交通道路和墙面。工作岗位照明可以由学生单独控制，将照明点锁定在需要的区域。

The existing buildings and new facilities are fully open, to improve the building. Designed to take full advantage of the main hall of the height of the original bank, including a series of laminated, interspersed a number of temporary tables and chairs, so that not only meet the need for manual work, but also engage in a computer-aided design.

These spaces include the "Tank", "Pool", "Wave". One, "Tank" is a custom blue glass box, placed in the embedded post. A fluid-filled "pipeline filling", the formation of a blue neon tube, wrapped in a glass box around the blue glass wall for the following additional fantasy glory. "Pool" is an elevated wooden platform, the platform has a large central depression, decorated with swim pad. This space is for informal discussion of the team place. Rest next to the custom chair with adjustable armrest, in order to bring portable computers to the people who work provides a relaxing, similar to the swimming pool side lounge chair. "Wave" is an independent mezzanine, looks like floating in the air. Bulletin boards and a large square table to form a space, students can engage in a personal creative can also cooperate with each other.

Studio wall printed with cactus pattern by Marimekko fabric wrapped decorative plates, so that environmental noise can be reduced. First floor, custom low tatami mats on the stage covered with fake fur, which adds to the informal working environment of diversity. Lighting can illuminate the traffic roads and walls. Jobs illumination can be controlled individually by the students will be required lighting points, locked area.

↑非正式的工作环境的榻榻米 / Tatami informal working environment.

↑正式办公区 / A formal office.
↓游泳池旁的休闲椅 / Leisure chair beside the pool.

↑游泳池上方的灯罩 / Swimming pool at the top of the shade.
↓平面图 / Plan

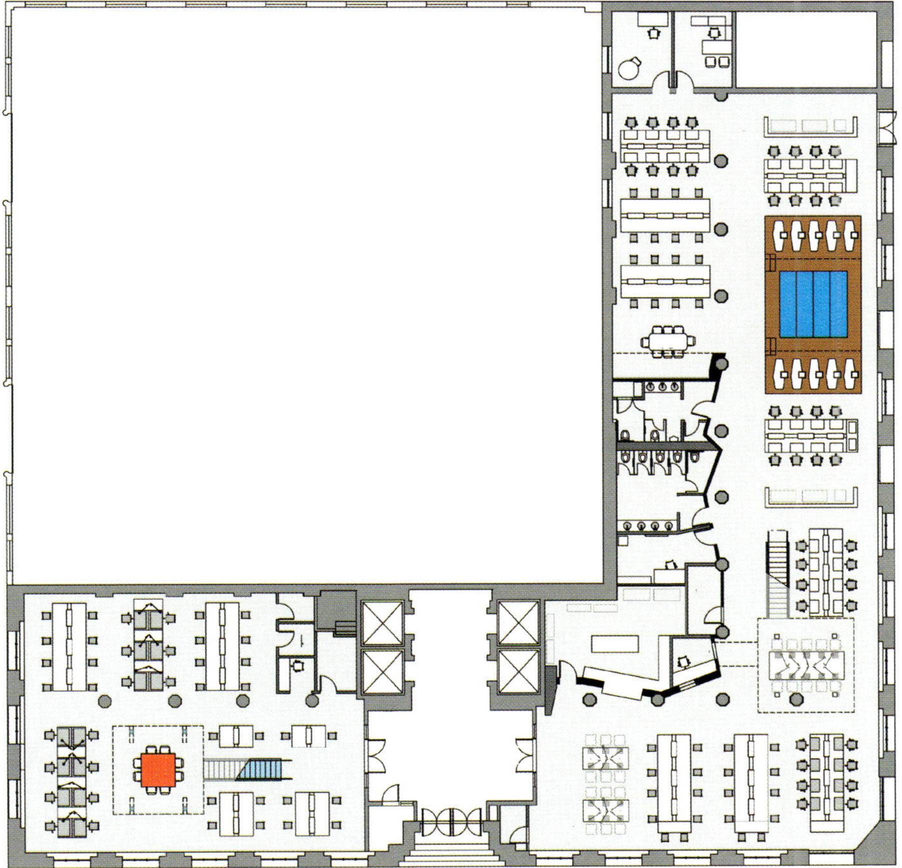

↑ “Pool”，“旱”的游泳池 / "Pool", "drought" in pool.
↓抬头看蓝色的“Tank” / Look up at the blue "Tank".

↑连接“Wave”的楼梯 / Connection "Wave" of the staircase.
↓从“Tank”向外望去 / From the "Tank" and looked out.

LVJING ZHENGXIE BUILDING

绿景政协大厦商务示范单位

34

【坐落地点】广东省深圳市
【面积】167 m^2
【设计】李益中
【主要材料】深灰色地毯、钢琴烤漆、金箔、皮革、铜镜

这个户型打造的是一个纯粹的办公空间，在意象投射中，设计师设想的是一个顶级品牌腕表生产商用于接待VIP客户的办公室，它要呈现的是与其品牌相对应的精湛工艺，优雅独特的造型以及整体散发出来的尊贵眩目感。

因此，设计师以极具现代感的金属金色为主色调，大量采用红铜镜面及灰色镜面不锈钢，在整体上营造一种冷冽，精细的氛围。同时巧妙的同皮革搭配，镜面与镜面，金属与皮革相互碰撞出一个有如钻石切割面般的空间，现代又不失高雅与尊贵。

会议室外设置了一整面产品展示墙，采用立体的钢琴烤漆方块，随意凹凸摆放，以白色为主色调，黑色点缀其中，黑白的强烈对比，凹凸的界面，映衬出所展示腕表的精细与不凡。展示墙对面设置了一个多功能的吧台，给来访的客户营造了一种轻松休闲的氛围。

The Unit is to create a pure office space, in the image projection, the designer is envisaged is a manufacturer of top-brand watches for the reception VIP client's office, it will be presented with their corresponding brand exquisite technology, elegant and unique distributed out of shape and the overall sense of the distinguished dazzling.

Thus, a very modern design with metallic gold-based colors, extensive use of copper and gray mirror mirror stainless steel, the cold on the whole to create a fine atmosphere. At the same time ingenious with the same leather, mirror and mirror, metal and leather like a collision between a diamond-cut surface-like space, modern without losing the elegance and dignity.

Outside the meeting room set up an entire wall of products, using three-dimensional piano paint box, free bump located, the main colors of white, black dotted among them, black and white contrast, concave-convex interface, to reflect a watch displayed Fine and uncommon. Show the wall opposite the bar set up a multi-functional, giving the visiting customers to create a relaxed casual atmosphere.

↑办公室 / Office

↑办公区 / Office ；↑卫生间 / Toilet
↓小型会议区 / Small meeting area.

↑办公会议区 / Office conference area.
↓平面图 / Plan

平面布置图 PRIMEVAL PLAN
SCALE: 1/75

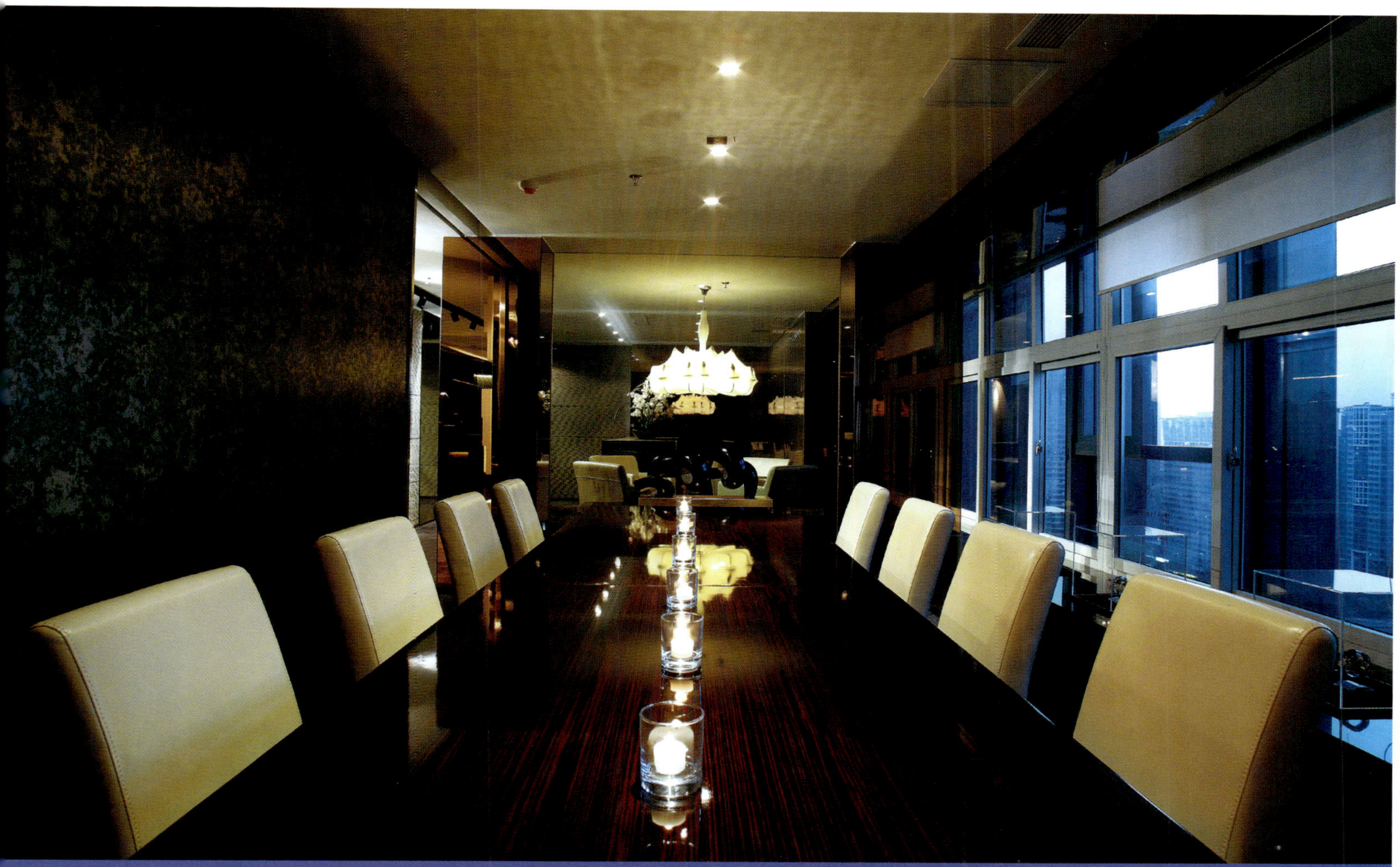

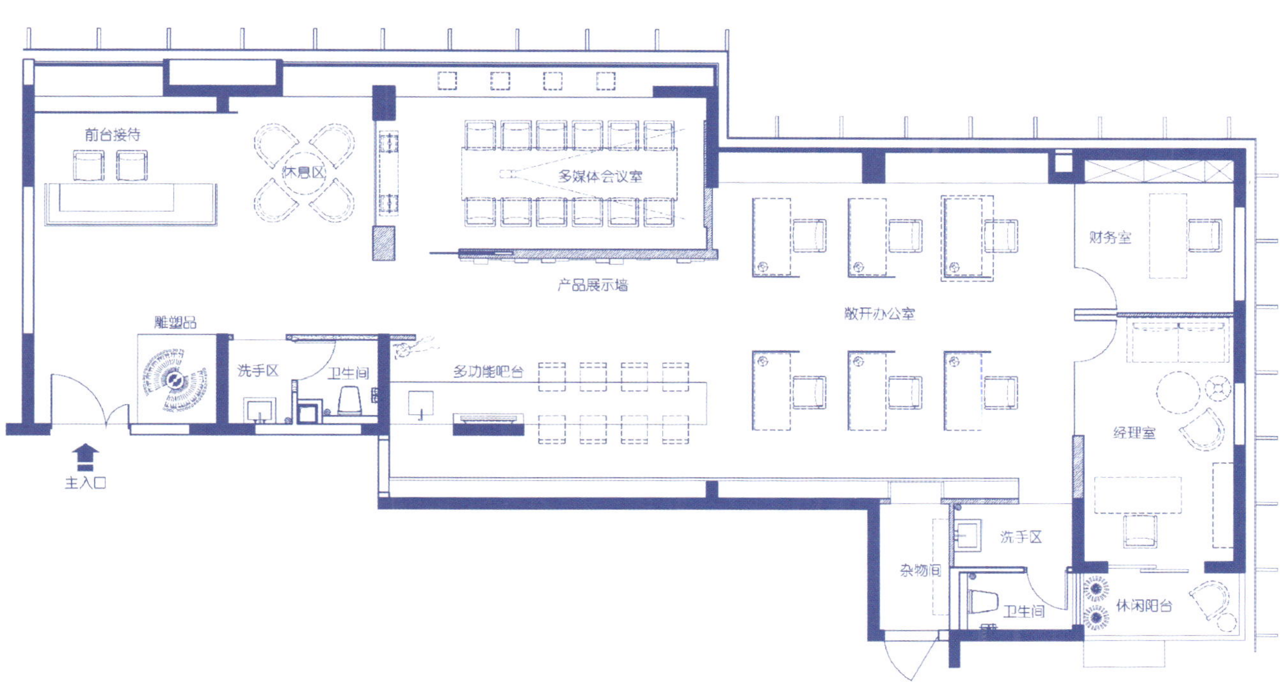

CUBION A/S OFFICE

Cubion a/s办公空间

35

【坐落地点】丹麦哥本哈根
【面积】178 m^2
【设计】Jackie-B 〔丹〕
【摄影】Jacob Nielsen

顾问公司坐落在一座文物保护建筑的第一层。整个公司办公空间有三个不同的部分：工作区、交流区、洽谈区。

在这个可以容纳8名雇员的办公空间里，淡雅的乳白色配合鲜亮的苹果绿，和引入墙壁的人造草皮，让在这里工作人也能放松心情。为了配合这样的格调，设计师还专门为此办公室设计了桌椅和台灯，营造出一种和谐的美感。人造草皮还有一个重要的功能，就是方便人们斜靠，绿色使人保持冷静，并且给人以美好的幻想。白色的书写板和透明的玻璃墙在保证了办公空间必要的私密性的同时，也供大家在头脑风暴的时候使用。

会议室的另一个功能是可以作为一个小型的厨房使用，明黄色调有助于促进食欲。临窗的小型吧台，丰富了的午后时光，是坐在窗边悠闲地看风景，还是借着温暖的日光懒散的阅读，由你选择。隐藏在复印机后面的小木屋，给人们提供一个可以沉思和放松的地方，一个相互交流沟通的平台。

Consultancy firm is located in a heritage building's first floor. Office space for the entire company has three different parts: the work area, communication area, to discuss area.

In this can accommodate eight employees, office space, the elegant milky white with bright apple green, and the introduction of artificial turf walls, so that people who work here can relax. In order to facilitate this kind of style, designers also specially designed office furniture and lamps, creating a sense of harmony and beauty. Artificial turf there is an important function, that is easy for people leaning, green people to remain calm, and gives a wonderful illusion. White writing board and a transparent glass wall of the office space needed in ensuring privacy while also brainstorming for everyone at the time of use.

Another feature of the conference rooms can be used as a small kitchen, bright yellow tone help to promote appetite. The window of a small bar and enriched the afternoon time, was sitting by the window leisurely look at the scenery, or the warmth of the sun through lazy reading, by your choice. Hiding behind the photocopiers cottage, you can offer people a place of reflection and relaxation, a platform to communicate with each other.

↑ 白色与黄色的搭配简洁大方 / White and yellow with simple and generous.

↑开敞明亮的办公区 / Bright open office area.
←休闲小吧台 / Leisure mini-bar.
↓↓充满浪漫气息的小餐厅 / Romantic little restaurant.

IPPOLITO FLEITZ GROUP

Ippolito Fleitz Group工作室

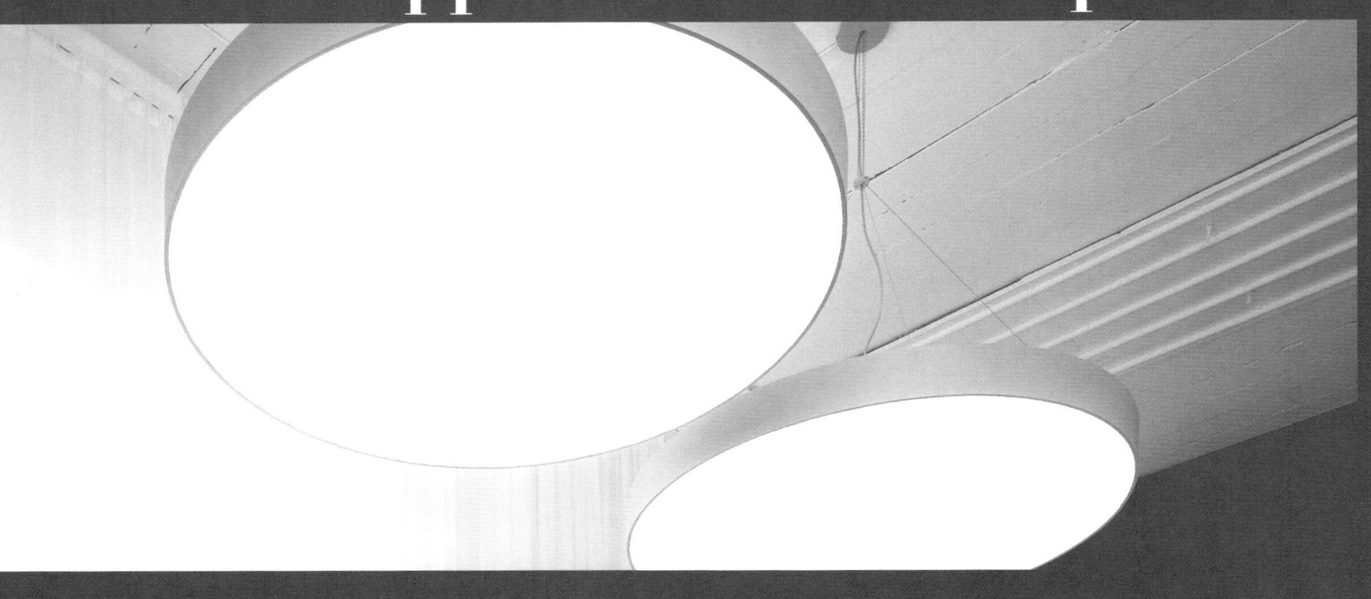

36

【坐落地点】德国斯图加特
【面积】480 m^2
【设计】Ippolito Fleitz Group设计工作室〔德〕
【摄影】Zooey Braun

工作室使用了大面积的白色作为主色调，同时在较小的空间内使用黑色和灰色块面来降低白色带来的高调和夺目。当然大的主色调中间也不时穿插一些明亮的色彩，用以调节空间可能出现的沉闷感。从室内陈设的选用来看，家具的选择也符合空间简约的设计风格，工作区的家具中规中矩，非常符合功能性要求。黑色的家具和白色的主调形成对比，相互调节，形成一种庄重、严肃的工作氛围。两个长桌子便于员工之间的齐心协力的交流和创作，着力营造一种平等、融洽、流畅的气氛。精心挑选的带状的纺织品装饰物鲜艳多彩，仔细观察，上面还印有漂亮的几何纹样。

休闲区和会议区的家具则精心挑选，颇有特色。设计师在此使用了白色的家具与原本雪白的墙体连成一体，带来一种别样的视觉体验。在这片纯白的空间中，材料间质感的对比越显突出，白色纱帘给室内带来动感的气息，而格子图案的地板的使用匠心独具，让人们在迷离的白色中寻找到一丝沉稳与安定。洽谈区的黑色墙面上悬挂着ippolito fleitz group过往作品的照片，让客户交谈之余可以对他们的设计窥见一斑。

Studio using a large area of white as the main colors, while the use of space in smaller black and gray blocks to reduce the white surface caused by high-profile and eye-catching. Of course, the middle of a large main colors interspersed from time to time some bright colors, to adjust the dull sense of space that may arise. The choice of interior furnishings from the point of view, the choice of furniture, also in line with space-simple design style, work area furniture from law-abiding, very consistent with functional requirements. Black and white furniture, the main theme of contrast, mutual adjustment, to form a solemn, serious working environment. Two long tables together to facilitate communication between employees and creative, and strive to create an equal, harmonious and smooth atmosphere. Carefully selected band of textile decoration bright colorful, carefully observe the above geometry is also printed with beautiful patterns.

Recreation area and meeting areas of the furniture is carefully selected, rather special. Designers in this use of white furniture, white walls fused with the original, bringing a different kind of visual experience. In this piece of pure white space, materials, texture contrast between the more conspicuous white gauze to bring dynamism to the indoor atmosphere, while the use of plaid creative design of the floor, so that people find in the white blurred trace calm and stability. Discussion area on the black wall hung ippolito fleitz group photos of past works, so that customers were chatting I can get a glimpse of their designs.

↑接待区装饰简单大方 / Reception area decorated simple and generous.

↑庄严的办公区 / A solemn office.
← 圆形的大吊灯 / Circular chandeliers.
↓平面图 / Plan

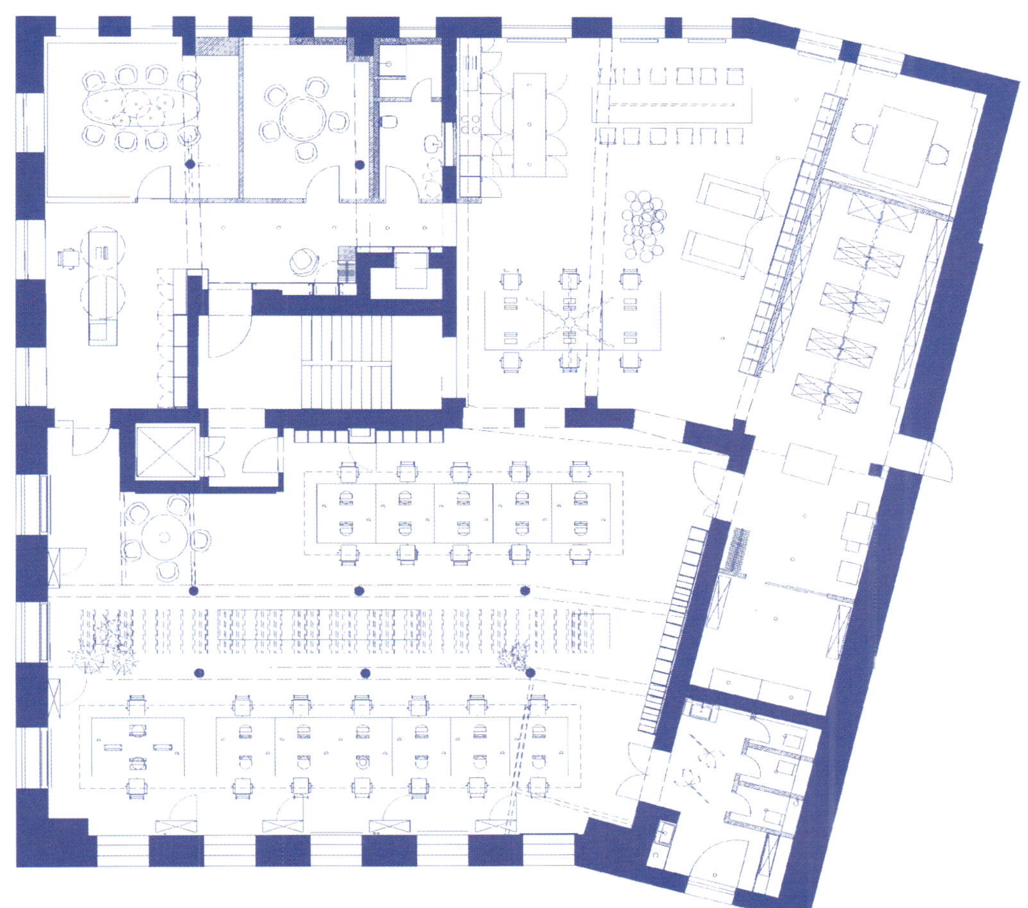

↑巨大的书架陈列 / A huge display shelves；↑员工学习区 / Their staff in the area.
← 会议室 / Conference room.
↓办公空间 / Office space.

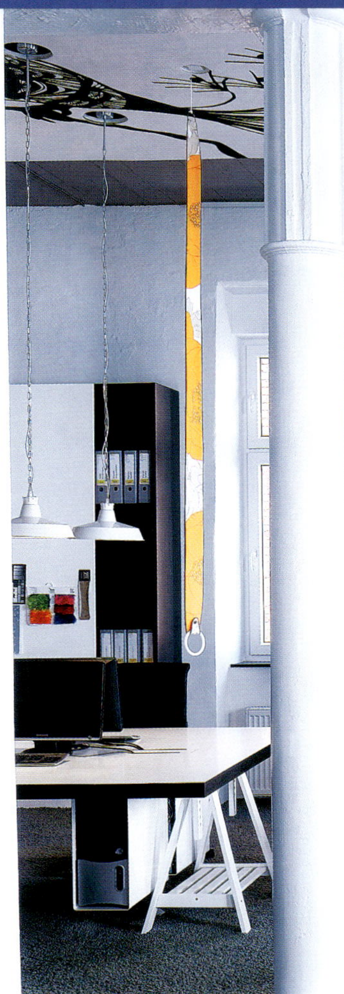

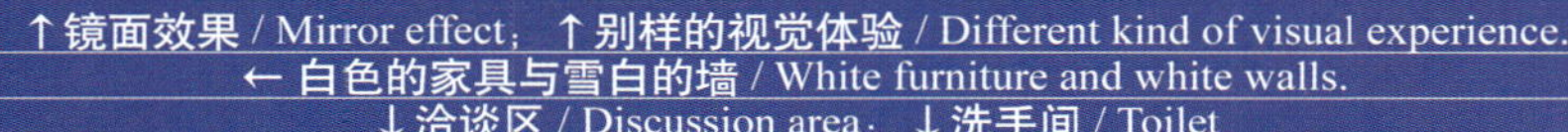

↑镜面效果 / Mirror effect；↑别样的视觉体验 / Different kind of visual experience.
←白色的家具与雪白的墙 / White furniture and white walls.
↓洽谈区 / Discussion area；↓洗手间 / Toilet

特别推荐 HIGHLY RECOMMENDED

顶级设计空间Ⅰ——情调餐厅

ISBN 978-7-5038-5798-0
印装：四色精装
定价：248.00

顶级设计空间Ⅱ——纯粹商店

ISBN 978-7-5038-5797-3
印装：四色精装
定价：248.00

顶级设计空间Ⅲ——创意办公

ISBN 978-7-5038-5803-1
印装：四色精装
定价：248.00

顶级设计空间Ⅳ——奢华酒店

ISBN 978-7-5038-5802-4
印装：四色精装
定价：248.00

住宅字典
住宅立面造型分类图集. *1*
ISBN 978-7-5038-5756-0

住宅字典
住宅立面造型分类图集. *2*
ISBN 978-7-5038-5755-2

住宅字典
住宅立面造型分类图集. *3*
ISBN 978-7-5038-5754-6

香港日瀚国际文化传播有限公司编
出版：中国林业出版社
印装：四色平装
开本：218mm × 336mm
版次：2010年1月第1版
印次：2010年1月第1次
单册印张：20.25
单册定价：198.00

室内设计新作（上下卷）
张青萍 主编
ISBN 978-7-5038-5435-4
开本：230mm × 300mm
页码：700
印装：软精装
定价：558.00元
出版时间：2009年6月

顶级样板房Ⅰ
张青萍 孔新民 主编
ISBN 978-7-5038-5709-6
开 本：230mm × 300mm
页 码：360
印装：精装
定 价：288.00元
出版时间：2009年10月

联系单位：中国林业出版社
地址：北京西城区德内大街刘海胡同7号
邮编：100009
销售客服：13641384559
出版客服：13810400238

网络支持:www.onetopspace.com